THE
COMPLETE HANDBOOK OF
POULTRY-
KEEPING

THE
COMPLETE HANDBOOK OF
POULTRY-
KEEPING

STUART BANKS

VAN NOSTRAND REINHOLD COMPANY

NEW YORK CINCINNATI TORONTO LONDON MELBOURNE

© Stuart Banks 1979

First published in Great Britain 1979 by Ward Lock
Limited, a Pentos Company.

Library of Congress Catalog Card Number
79-14305

ISBN 0-442-23382-5 cloth
        0-442-23383-3 paper

Printed in Great Britain

Published in 1979 by Van Nostrand Reinhold
Company
A division of Litton Educational Publishing, Inc.
135 West 50th Street, New York, NY 10020,
U.S.A.

Van Nostrand Reinhold Limited
1410 Birchmount Road, Scarborough, Ontario M1P
2E7, Canada

16 15 14 13 12 11 10 9 8 7 6 5 4 3 2 1

**Library of Congress Cataloging in Publication
Data**
Banks, Stuart.
    The complete handbook of poultry keeping.

    Bibliography: p.
    Includes index.
    1. Poultry.   2. Chickens.   I. Skinner, John L.
II. Title.
SF487.B182   636.5′08   79-14305
ISBN 0-442-23382-5
ISBN 0-442-23383-3 pbk.

Acknowledgements

Special thanks are due to John Portsmouth for so
promptly and efficiently supplying the chapter on
nutrition for this book, and to Professor Keith
Skinner of the University of Wisconsin for reading
and suggesting amendments to the manuscript. I am
indebted to colleagues on *Poultry World* and many
friends in the industry for their help in supplying
information and support, and to my long-suffering
family, who began to believe it would never be
finished.

The author and publishers would also like to thank
the following for kindly supplying pictures for the
book:
Rank, Hovis McDougall Ltd, page 91; Smith, Kline
& French Laboratories Ltd, page 173; Mike Parker/
Poultry World, page 6; *Poultry World*, all other
pictures.

We would like to thank Peter Hand (GB) Ltd, for
the food analysis tables on pages 143–5 and the
Ministry of Agriculture for information given in the
table on page 111. The diagram on page 105 is
based on the *Ministry of Agriculture Bulletin 56*.

Frontispiece An example of a modern brown hybrid
layer, the Babcock B380

Contents

Foreword

The poultry world comprises some very versatile and extremely useful creatures. They are well adapted to serving man in a variety of ways. This service ranges from the strictly economic consideration of efficient food production to the purely esthetic attributes of color and form. Stuart Banks has managed in *The Complete Handbook of Poultry-Keeping* to cover in depth the broad expanses of this industry.

Because of Mr Banks' deep personal interest and years of professional experience in the poultry community, he has been able to present in this book a unique amount of poultry understanding. He very effectively lays the groundwork of considerations for the inexperienced person who is about to try his (or her) hand at poultry production. He also deals with the economic considerations that form the basis upon which large commercial producers decide to expand. This truly reflects the author's understanding of the total poultry picture.

This volume treats the subject of poultry in a truly universal way. Mr Banks weaves the results of research and commercial production together and shows that these factors are some of the main points of concern for the backyard flock owner and the fancier as well. This is particularly timely with the renewed interest in home food production. It is also fortunate because it comes at a time when many small operators and home food producers are expressing the feeling that the results of research and the forces of commercial production are working against them.

Combining as he does a bit of history and a touch of forecasting with a sound understanding of the current 'state of the art', Stuart Banks has created an interesting and factual treatise. It will serve as fascinating reading for the uninformed. At the same time *The Complete Handbook of Poultry-Keeping* can be a ready reference for students and beginners while concurrently offering the background information that commercial operators need to reconsider from time to time.

John L. Skinner
Poultry and Small Animal Specialist
University of Wisconsin

1 First decisions

One of the great attractions of poultry-keeping has always been its accessibility. A bird is compact, does not eat much, does not, normally, cost very much to buy. It can be kept as a pet, a hobby or a business proposition and management, on the small scale at least, is relatively easy. Little wonder then that poultry is seen by many as an ideal step towards backyard 'farming' and some degree of self-sufficiency.

There is a danger in this apparent simplicity, however. It tempts people to drift into poultry raising with little or no thought about their real objectives. A vague notion about having home-produced eggs is translated into buying some chickens. Then the questions arise. When will they start to lay? How many eggs in a week? If I keep brown and white birds together, will they fight? How long should I keep layers? All perfectly valid questions, but rather late when your new purchases are clucking in the back of the car! Far better to think first and apply some method to your planning.

Poultry is a very wide subject and you might even find, on consideration, that egg production is not your main interest after all. Show birds, for instance, have a tremendous following and need not take up any more space than layers. In fact, if you keep bantams they will take less. The standard breeds have a fascination of their own. Then there are the other species. Ducks are generally more prolific egg layers than chickens, good table birds, easy to manage. Others in the meat category are turkeys, geese, quail and guinea-fowl.

Your choice is broad, therefore, but whatever line you decide to adopt, some knowledge of the professional side of the picture will also be of value to you. Apart from the possibility that you may have ambitions to enter commercial production, the modern poultry industry has a profound influence on producer and consumer attitudes. In making eggs and poultry meat more widely available than at any time in the past, it has introduced new management methods, new veterinary aids and high-performance hybrid birds which cannot be ignored even by the domestic poultry-keeper. On its way to today's multi-million bird companies, only possible through intensive farming techniques, the industry has attracted its share of criticism. Some has been warranted but a good deal is misinformed. Certainly poultry products would be put firmly back in the luxury bracket by any return to the old, traditional free-range farming.

Making a start

From the beginning you have to be sure about two basic essentials, time and space. Remember that, however few birds you keep, somebody has to find time to feed them (and if you feed scraps, that includes preparation), water them, clean out their living quarters, if they are layers collect their eggs and give them a daily talking to. Strange as it may seem, even in the largest battery houses, the best stockmen are almost invariably those who have a 'feel' for their birds and talk to them. It shows in the results.

Unless you are planning a totally single-handed operation it is wise to bring the family into the debate early in your planning. None of the activities described so far is going to take more than a few minutes, but it is as well for everybody to understand that it is going to cost some time every day, and if you cannot do it somebody else must be prepared to step in. Holiday stand-ins are worth a thought too. I know of a commercial producer who could pinpoint his holidays by the dip in his egg graph. Birds need conscientious attention at all times.

Does the family agree about the size and scope of the poultry project, and is everybody healthy? Bear in mind that feather dust and the atmosphere around poultry can set up reactions in some sensitive people. Make sure that nobody likely to come in contact with the birds has an allergy to them.

Have you the space? This depends on your circumstances, the species you want to keep and in what numbers, but chickens do not require a lot of room. A 'run' is not essential, even for poultry kept domestically. I have on my desk, as I write, a report of over 100 dozen eggs being produced in a year by four hybrid hens kept in cages in a garden shed 2.3 m × 1.7 m (7 ft 6 in × 5 ft 6 in).

As many as thirty light hybrids may be kept in a shed 1.5 m × 1.8 m (5 ft × 6 ft) with an adjoining run about twice that area, but a great deal depends on the discretion of the poultryman. Heavy breeds, when full grown, can exceed 4.5 kg (10 lb) in weight and would obviously be hopelessly cramped in quarters which would comfortably house growing birds or modern lightweight breeds.

Ducks, like chickens, are not particularly space-demanding. A typical domestic group will consist of a drake and six or seven ducks, and they do not need a pond, although they will enjoy one if it can be provided. They will live quite happily in the open, preferably penned on a grassy area, but to protect them—and their eggs—from predators and the worst extremes of the weather they can be provided with a simple shelter or ark. This need be no more than 90 cm (3 ft) high and 2.1 m × 1.2 m (7 ft × 4 ft) in area.

Geese are a different proposition, with an appetite for grass. It has been estimated that five geese will eat as much grass as one sheep, so only if you have access to a suitable field, or the sort of lawn to feed a sheep or two, should you consider this species.

Turkeys, like geese, are large birds best suited to farm conditions or the facilities of specialist growers. Traditionally they are kept in pole-barns (open-sided sheds), but industrial producers are moving progressively to completely controlled-environment buildings. Generally a number of birds are needed to make an economical crop and under the UK Animal Welfare Codes of Practice recommendations a bird weight of 24.4 kg should be allowed for each square metre (5 lb per ft^2). Three main weight ranges have been developed to meet market requirements in the UK and elsewhere, 'mini', 'midi' and 'maxi' birds maturing at different sizes. These are sold at liveweights from 3.2 kg to about 16 kg (7 lb–35 lb+). Reared in the open, turkeys are stocked at about 400 to the acre or nearly 1,000 to the hectare.

Individually, quail and guinea-fowl present no space problem, but you need a fair number of these small birds, of course, to make a business. Quail can be kept in floor pens or cages of the sort used for battery-brooding of chickens. Guinea-fowl are run outside, where 500–700 birds to the acre (1,200–1,700 per hectare) may be kept on good range. In-doors, producers allow 6–9 cm^2 ($\frac{3}{4}$–1 ft^2) per bird.

With the family on your side and a fair idea of the time and space you will need, the next stage is to see whether the neighbours are also going to see things your way.

Ill-managed poultry can be noisy, smelly and an attraction for vermin. They can also stray on to other people's land if not properly supervised. All these transgressions can bring the poultry-keeper up against outsiders, so not unnaturally a variety of regulations and laws are available to control the situation.

The most immediate question is whether poultry-keeping is permitted at all. Quite frequently, particularly on estates, covenants are laid down governing the use or purpose to which property can be put. There may be a ban on the keeping of pigeons say, or pigs, and just possibly poultry. For owners the facts can be established by inspecting the property deeds. Leaseholders will find them in the terms of their lease.

The next line of inquiry is to the environmental health department of the local authority. Some authorities apply by-laws, but the majority in the UK allow domestic poultry-keeping and rely on the Public Health Act to control any nuisance. Among its many provisions, for example, this Act empowers local authorities to serve an abatement notice on anyone keeping an animal 'in such a place or manner as to be prejudicial to health or a nuisance'. In default, a fine can be imposed for each day the nuisance continues, plus the council's costs in putting the matter right.

Assuming that official permission is no problem, however, it still makes sense to keep on good terms with the neighbours. An offer of eggs or some other share of the produce, when it comes, might be welcome, but the most obvious duty is to ensure that the birds cannot encroach on neighbouring land and do enough damage to destroy your goodwill.

The laws of trespass would allow damages to be claimed where, for instance, a claimant could show that damage had been caused on his property by a defendant's poultry due to negligence in fencing them in, and prove that his costing of the damage was accurate. If the defendant could show that every reasonable precaution had been taken and that some third person's carelessness was responsible, the claimant's case would fail.

Additional buildings or alterations to existing ones raise another aspect of 'permission'. Normally, for the type of installation required by the domestic poultry-keeper, which as we have seen need amount to no more than a garden shed, no special planning permission is required.

If a proposed building has a volume of more than $30\,m^3$ ($1,059\,ft^3$) or comes within two metres of the house, it does become subject to building regulations. Be on your guard also for anything which may step outside the building line or entail widening a drive-way on to a trunk or classified road. And remember that planning permission is required for professional or business premises.

Costs

You are going to need livestock, somewhere to keep them, food and water at the outset. If you are going to hatch and rear birds, either now or later, you will also need some form of incubator and heat sources. These, with your own valuable time, are the main costs incurred in any poultry enterprise. Basically it is the cheapest form of productive livestock, within the reach of most pockets, and quite a lot of children take great pride in units they have set up and stocked themselves. The one qualification is that pure breeds, for obvious reasons, are likely to be expensive if they are good show birds.

Feed is the biggest single running cost, representing around 70% of a commercial poultryman's outlay in the UK (somewhat less in America). On a small scale this is less of a worry, because often the poultry ration can be supplemented by domestic food scraps and greenstuffs and the actual output of eggs or meat-gain per bird is not so crucial. Also the limited quantities needed mean that the cost is not a great burden. In effect, the small flock owner always pays more for any commerical ration, weight for weight, than the professional, of course, because he is buying in 25 or 50 kg ($\frac{1}{2}$ or 1 cwt) bags, without any of the advantages of bulk purchase and handling. A 25 kg ($\frac{1}{2}$ cwt) bag of feed will last six pullets a little over a month.

So many are the variables that it is not feasible to quote actual feed costs. In one year, for example, the average cost of mixed feeds for

UK egg producers rose by over 30%. Around this average some producers would have paid more per tonne because they were buying small amounts. Others, generally the larger producers, may have bought at favourable prices on the forward market, paid less by buying in bulk and possibly had some acreage of their own grain. Some producers save by having their own milling and mixing plant to prepare their rations.

Feed may be medicated. It can have different ingredients. It can vary in delivery price according to how close you are to the mill. The permutations are innumerable. The only satisfactory way to find out what price is right for you is to make direct inquiries to the suppliers and wherever possible obtain quotations from more than one source.

Sources of supply

Naturally, when it comes to buying livestock, the source depends on the species. Commercial hatcheries of any size these days deal almost exclusively in the modern hybrids produced by the major breeders of laying and meat stocks. A summary of these, and the sort of performance to be expected from them, appears in Chapter 3. Most hatcheries will sell small batches of chicks, but they cannot always undertake the delivery of 'penny numbers' and if you want less than, say, 200, you may have to arrange to pick them up.

'Day-olds' are a common starting point with poultry, but rearing brings its own requirements. The birds must be kept warm, fed, watered and lit correctly. To do the job properly, they should also be vaccinated according to an approved programme. As a result quite a lot of poultrymen, including some professionals, prefer to buy laying stock at 'point-of-lay' from specialist growers. You pay more per bird, of course, but providing they have been properly grown, the chances of losses are less. These birds are fully feathered, of about twenty weeks of age and practically full grown.

Since travelling is stressful, especially to adult birds, it is advisable, other things being equal, to buy wherever possible from your nearest source. Suppliers frequently advertise in local newspapers, and the *Yellow Pages* list livestock breeders and poultry farmers. It is a good idea to look at the poultry magazines as well. Apart from advertisements, they frequently publish directories and supplements detailing companies in the industry, with the bonus that they are a useful means of keeping in touch with latest developments in the industry.

All the sources mentioned so far are equally valuable for finding feed and equipment suppliers, but what if you want to run a unit on traditional lines, with the old strains and crosses no longer considered competitive by commercial producers—have they totally disappeared? This question is still asked, particularly by old-timers returning to the industry, and the answer is no, not completely. The Rhode Island Reds, the Barnevelders and Marans, the White Leghorns and Light Sussex, along with many other breeds, have been maintained by dedicated breeders, though in nothing like the numbers they used to be, and they are still obtainable. Your best reference here is to the various poultry clubs, who have records of the breeders of both full size and bantam poultry (see Chapter 4, page 39).

Other options

Outside the industrial applications of egg and meat production described in Chapters 13 and 14, there is no shortage of other activities for the enterprising poultryman. I have touched on the fascination of breeding birds for show, and this is dealt with in more detail in Chapter 4, but what about eggs for show? They too can win prizes on display. Some hens lay eggs of unusual colour. The Araucana, for example, has become well known for its beautiful, blue-shelled eggs, and it may be reared for these alone, to give novel effect to a competition entry.

Feathers, in themselves, may hardly be considered a reason for keeping poultry, but modern conditions have created a situation where it can pay to think carefully about the type and quality of feather your stock are going to yield.

In general, feathers tend to be roughly treated in commercial processes these days. Scalding and wet plucking do nothing for the quality of broiler and turkey feathers, which cannot be salvaged in their whole form, although they are still worth something when cooked and ground to make feather-meal—a feed ingredient—or as a fertilizer. Wax plucking similarly spoils the feathers of ducks and geese.

These wide-scale practices open up possibilities for the small or specialist producer, because the demand for feathers in good condition is, if anything, rising. The drawback is in producing them in sufficient quantity to be commercially realistic, but where an outlet is found, it should not be beyond the poultryman's wit to exploit it, possibly in co-operation with other producers.

Requirements range from the down and soft feathers, sought even now by certain traditionalist bedding and furniture manufacturers, to the large hard feathers employed as flights for darts. Goose quills for arrows are imported to Britain from China because of the shortage on the home market.

The colourful feathers of some breeds are in constant demand for angling flies. It used to be popular sport, and probably still is, to filch plumes from the cages of entries at poultry shows.

Hat trimming is a conventional use for feathers which can be extended to the making of complete hats, brooches and other items of clothing. I have even seen bikinis—attractive if impractical—made of feathers.

So the opportunities for the poultry-keeper are legion, and certainly need not stop at egg or meat production.

2 Physiology

The hen is not an egg-laying machine, though performance with modern genetic refinements might tempt some to think so. No handbook is supplied with poultry saying which switches to press for a given reaction, but every producer should at least have a working knowledge of the highly complex organic structure of birds. Allowing for the obvious variations between species, the basic pattern is the same for all.

They share certain characteristics with us; warm blood, dependence on air, food and water, sensitivity to pain, but birds are not human. In some ways they may be considered superior to us. If a woman were to duplicate a pullet's performance she would need to produce babies equivalent to about ten times her own bodyweight in a year.

The distinction was neatly made by Dr A. B. Gilbert, of the ARC Poultry Research Centre, Edinburgh, in 1970. He compared an average laying hen of 2 kg ($4\frac{1}{2}$ lb) bodyweight with an average woman weighing 60 kg (132 lb) or about thirty times as much as the hen.

During a nine-month pregnancy a woman deposits, in the foetus, material equivalent to approximately 6% of her bodyweight. The hen accomplishes the same thing in two days, by laying just two 60 g (2 oz) eggs, and continues the process for long periods.

The efficiency of this reproductive process is quite staggering. Even to obtain such a bulk of material from the diet is no mean feat. A laying hen eats approximately 120 g ($4\frac{1}{4}$ oz) of food per day, providing about 18 g ($\frac{1}{2}$ oz) protein, 4 g ($\frac{1}{10}$ oz) fat, 60 g (2 oz) carbohydrate and 10 g ($\frac{1}{3}$ oz) minerals, of which the major part is calcium.

Dr Gilbert calculated that the equivalent human dietary intake would need to be over 500 g ($17\frac{1}{2}$ oz) of protein, 120 g ($4\frac{1}{4}$ oz) fat and nearly 2 kg ($4\frac{1}{2}$ lb) of carbohydrate daily. But since the foodstuffs for human consumption have not been so completely processed, the woman's total intake would have to be higher. Meat, for example, only has a 20–25% protein content and potatoes contain only 20% carbohydrates, so a woman would need to eat about 2 kg ($4\frac{1}{2}$ lb) of meat, 2 kg ($4\frac{1}{2}$ lb) of potatoes and 2 kg ($4\frac{1}{2}$ lb) of cereal or beans.

She would still be severely deficient in calcium and would need to eat about thirty teaspoonfuls of pure chalk or 10 kg (22 lb) of cheese a day—roughly twice the average annual cheese consumption per head in Britain!

Typical of today's commercial light hybrid layer is the Shaver 288

The calorific value of such a diet is 13,000 Kcals per day; even a man doing very heavy work requires less than 5,000, which is a fair indication of the effort of the hen, but also underlines the great difference between the bird's metabolism and the mammal's.

Feathers

The most obvious distinction between birds and the rest of the animal kingdom is in their feathers. These unique structures are a remarkable example of natural adaptation to circumstances, since they are believed to have evolved from the scales of the bird's reptile-like ancestors. Scales, of course, still exist on the legs of birds.

Feathers are of three basic types: contour feathers, known technically as *penna*; down or *semi-plume*, and hair-like feathers, *filoplume*, these being the 'hairs' singed from table chickens and turkeys. They are composed mainly of a substance called keratin and are set in follicles of epidermal tissue.

Structurally, main feathers consist of a cylindrical quill, or *calamus* leading into a shaft, the *rhachis*, from which grow the *barbs* which give the feather its shape. Each barb has two rows of *barbules*, and in the case of flight feathers, the barbules in turn have tiny hooks or *hamuli* which hold them together and provide wind resistance.

Arrangement of the plumes follows a set pattern of 'tracts' or *pterylae* on the body, separated by areas of skin without follicles, the *apteria*.

In general, birds lose and replace all their feathers annually, in one or two moults. A full moult tends to be associated with a cessation of egg laying—though occasionally a bird will lay throughout its moult—and under natural conditions poultry moult in the autumn. It can last from two to three months in chickens, and high egg producers are noted for moulting more rapidly than low ones.

Moulting is a delicately balanced process controlled by hormone secretion, and sudden disturbances of routine, particularly in nutrition or lighting, can bring it about. Sometimes the poultryman will turn this to his advantage by inducing a moult when it suits him. But the important thing is not to start a moult unwittingly by clumsy management—by forgetting to provide water regularly for example.

Feathers make up about 5% of bodyweight and a bird does not go through the process of replacement without a certain amount of trauma.

As a warmth-preserving, weather-proof coating and a protection against injury, feathers are highly efficient, and they are vital aids to flight, of course. An interesting experiment reported in 1976 by Dr Stuart Richards of Wye College showed how dependent hens are upon them.

Poorly-feathered birds, with nakedness over about 65% of the body surface, were compared with well-feathered hens, and they were found to produce up to 45% more metabolic heat in an effort to offset their 'baldness', when surrounding air temperature was about 15–20° C (59–68° F).

No matter how defective the plumage, the organs within the body must be maintained at around 41° C (105.8° F) to perform normally. But unlike many mammals, avian species do not have the ability to reduce blood flow near the skin in cold conditions—except in the relatively small exposed areas of the wattles and feet—so heat loss without feathers is rapid.

Poor feathering in the case mentioned, therefore, would mean that the bird would have to compensate either by eating around 40% more, or dropping egg production, and possibly both.

Skin

Although the skin of birds is similar to that of mammals it is a much thinner covering, fairly loosely attached except at the beak and legs. It has no sweat glands. This does not mean that no evaporation takes place through the skin. Recent research indicates, in fact, that more than half the water evaporated from the body of the fowl in air temperatures of 10–20° C (50–68° F) escapes via the skin. But it does mean that this cooling effect is not controlled by bodily reaction.

Panting is the normal method of shedding heat, and it is something the stockman should watch for and remedy when conditions are hot.

Combs and wattles are the only prominent areas of skin to be seen on chickens, and they have a purpose in dissipating heat, much like the radiator of a car. They contain innumerable small blood vessels lying close to the surface. Removal of the comb and wattles—which can be achieved by a simple, bloodless operation with scissors at day-old—has occasionally been advocated as a way of preventing more stressful damage later, and of reducing heat loss in cold weather. Experiments have indicated that egg yields have been improved with 'dubbed' birds. But quite apart from the disapproval of welfare authorities, the actual increases in egg output, (2% or less) do not justify the labour involved.

Development of the comb varies according to the presence of male or female sex hormones, and is quite easily changed by the injection of opposing hormones, as in caponization.

In the non-laying hen, also due to the change in secretion of gonadal (sex) hormones, the comb becomes soft, small and pale in colour, so it can be an indicator of laying condition. The cock's comb appears in a wide variety of shapes and sizes according to breed, but it is generally larger than in the hen.

The turkey has caruncles, fleshy protuberances on the head and neck and a snood, again more prominent in the adult male than the female, and similar skin features are seen on the face and over the base of the bill in Muscovy ducks and drakes.

Only one gland has an external opening on the skin in birds. This is the *uropygial* or preen gland situated at the base of the tail. When stimulated by pressure from the bird's beak, this releases an oily secretion which is spread on the feathers.

Skeleton

Since it is built for flight, the avian skeleton has two immediately noticeable features, streamlining and rigidity. Two broad plates of bone, the pelvic girdle at the top and the breast-bone or sternum beneath, provide a strong, tapering framework for powerful muscles, and protection against shock and impact for internal organs. By mammalian standards there are relatively few mobile joints, but that is not to say that birds lack essential mobility. If you doubt this, try turning your own head through 180°.

Most schoolchildren know the other property of bird bones, which is that many are hollow, to aid flight. They are actually pneumatic, containing extensions of the special air sacs which lead out of the lungs. Thus some of the vertebrae, the lumbo-sacral mass, the pelvic girdle, the first two vertebral ribs, part of the sternum, the humerus and the distal half of the coracoid (running between shoulder and sternum), have air pumped into them. This both lightens the structure and possibly adds to the respiratory effectiveness of the bird when using its highly developed muscles. Elsewhere, bone marrow performs the same function as in mammals, producing blood cells.

There is a specialized form of bone, important to the laying hen, known as medullary bone, which acts as a calcium store. It is deposited in those bones which have an efficient blood supply. During active shell formation, the hen, which has, as we have seen, a very high calcium requirement, can call on this 'reservoir' of bone calcium if she is not receiving enough by intestinal absorption from her diet. It is not policy for the producer to rely on her ability in this direction however. Any prolonged imbalance of calcium and phosphorus can be drastic for the layer.

Bone problems can arise from dietary deficiency, showing up in the form of brittleness or leg weaknesses and bowing, particularly in heavy meat strains. Lack of Vitamin D_3 in the rations of birds without access to sunlight can lead to rickets for example, resulting in soft bones in chicks and temporary paralysis in laying birds.

Cod liver oil is a source of Vitamin D_3 and either limestone or oyster shell grits supply calcium. But as can be seen in Chapter 10, commercial rations generally take these requirements into account nowadays.

Circulation

Blood has the same purpose in birds as in mammals, to equalize temperature, circulate oxygen and nutrients to the body, and transport impurities away, but it operates at a higher temperature, propelled by a higher pulse rate.

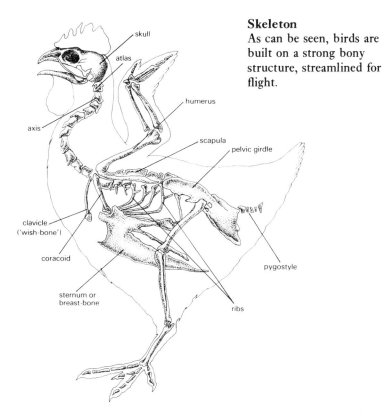

skull

atlas

axis

humerus

scapula

pelvic girdle

clavicle
('wish-bone')

coracoid

pygostyle

sternum or
breast-bone

ribs

Skeleton
As can be seen, birds are
built on a strong bony
structure, streamlined for
flight.

Body temperature in the healthy fowl varies, depending on the time of day, from a high of 43° C (109.4° F) in the afternoon to a low of 40.5 ° C (105° F) at around midnight.

The heart is well developed and commonly beats at well over 200 per minute even when chickens are at rest. Broadly speaking, the larger the bird the slower the pulse rate, so while a light white hybrid hen may have a resting heartbeat of 330 per minute, a cockerel of a heavy strain may average under 250 per minute.

Birds are not immune to heart attack. Heavy mature turkeys, particularly if over-fat and over-excited, are subject to aortic rupture—bursting of a large artery from the heart, generally where it branches into a right and left coronary artery.

Red blood cells, *erythrocytes*, are the oxygen carriers. They contain *haemoglobin* which becomes *oxyhaemoglobin* by absorbing oxygen in the blood vessels of the lungs, then reverts to haemoglobin by giving it up to the tissue cells.

White blood cells are *leucocytes* which can be divided into *lymphocytes* formed in the lymphatic organs and *granulocytes* derived from bone marrow. There are further divisions within these categories, but basically leucocytes are the body's defence against infection. They attack potentially harmful micro-organisms, remove particles, and have a special function in forming and distributing antibodies.

The *Bursa of Fabricius*, situated just inside the body and above the vent, and the *thymus* in the neck, are the central lymphoid organs. The *spleen* also has large amounts of lymphoid tissue, and these are the sources of lymphocytes.

Digestive system

Contrary to some opinion, poultry do have a sense of taste, with glandular taste buds associated with the ducts of the salivary glands, but their idea of appetising flavours is clearly rather different from ours.

Food is transferred rapidly to the *oesophagus* or gullet, a narrow, muscular tube which can be widely distended to accept undigested chunks. The oesophagus widens into the crop, and there the food, lubricated on its passage down the throat, may be held for a time, depending on the state of the bird's appetite. If the gizzard is empty it may pass straight to that stage of digestion.

Quite often food will stay in the crop for several hours, being softened by the moisture secreted by glands lining the organ. It then moves to the glandular stomach or *proventriculus* which is a short, thick-walled organ lying above the liver and between the oesophagus and gizzard. This, with the gizzard, corresponds to the stomach in mammals. In the *proventriculus* the enzymes *pepsin* and *rennin*, and hydrochloric acid come into action, and the digestion proper begins.

From there the food is transferred to the gizzard or *ventriculus*—a bird's 'back teeth'. This muscular organ, usually about 5 cm (2 in) in diameter and 2.5 cm (1 in) thick in the fowl, has both its entrance from the proventriculus and its exit set into its upper surface. It is a cul-de-sac and occasionally, if a bird has had access to long grass, hard fibre or other indigestible objects, it can provide a traffic jam with the 'tail-back' extending to the crop and causing impaction.

Sometimes the bird will itself clear the obstruction by convulsive movements of the neck and crop. It may be possible to help it by manipulation and massage with the fingers, otherwise it is a small job for the vet, who will make an incision in the crop. Without attention the bird will die.

To grind effectively, the gizzard needs grit, which the bird ingests. Two forms of grit are available; soluble and insoluble. They can be purchased in mixed form, and in sizes to suit different types of bird, but are they necessary?

Soluble grits are those which bear calcium—limestone, oyster shell and now cockle shell are used. They provide layers with essential calcium but have obvious limitations for grinding, since they dissolve. Also most compound rations provide the calcium balance. Insoluble grits are chippings of stone such as granite which does not create sharp edges like flint or quartz, which might be dangerous to the bird. Modern milled rations are sometimes said to be sufficiently prepared to make the gizzard redundant.

Thoughts have begun to turn full circle on the subject of grits, however, and such authorities as Professor Milton Scott of Cornell Uni-

Digestive system

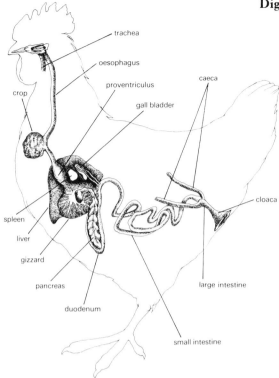

trachea
oesophagus
proventriculus
caeca
crop
gall bladder
spleen
liver
gizzard
pancreas
duodenum
large intestine
cloaca
small intestine

versity support the use of insoluble grit when, for example, oyster shell is used as an extra calcium source. He found that consumption of the soluble grit went down by two-thirds, indicating that more calcium was being released from it. Professor Scott also feels that birds may derive a 'sense of well-being' from grit, and tend to over-consume oyster shell when it is not provided.

My views are shaded in favour of giving grits, even to professionally caged layers, for these reasons and also because of their possible marginal improvement on digestive efficiency. Certainly, where whole grains are being fed, or home-produced rations with roughage, some access to grit is important. It also helps the gizzard to deal with the occasional feather that is swallowed.

Leading out of the gizzard is the *duodenum*, the upper part of the small intestine, which loops round the *pancreas*. Supported in the *mesentery*, which is a transparent membrane extending from the lining of the body cavity to hold the viscera in place, the pancreas is a long, narrow, grey-white gland. It has two types of tissue, the enzyme-secreting cells and the endocrine, hormone-producing cells.

The enzymes *trypsin*, *diastase* and *lipase* contained in the pancreatic juices are ducted to the duodenum. There, together with other enzymes secreted by the intestine, they hydrolyse proteins, carbohydrates and fats respectively, breaking them down into amino acids, simple sugars, and glycerine and fatty acids, which can be absorbed through the gut wall into the bloodstream.

The small intestine, much shorter, relatively speaking, than in mammals, is generously supplied with *villi*—like the pile on a towel—which give the epithelial lining a considerable surface area.

The liver is the second of the major glands contributing to digestive secretions and it is the largest of the abdominal organs. Since it uses materials derived from food and temporarily stores some, it is subject to dietary intake, particularly fat. This is why force-feeding geese increases their size and creates a 'fatty liver', for example. But another of its functions is to produce bile.

Bile travels from the liver through two ducts to the duodenum, either direct, or via the gall bladder, which concentrates it before passing it on. In the intestine it has the effect of emulsifying fats so that the digestive juices can act on them.

The final section of the alimentary canal is taken up with the large intestine and the cloaca. Two long, blind sections of gut, the *caeca*, open out of the junction between the small and large intestines. Although they fill with faecal matter and empty regularly they appear to have little bearing on digestion generally.

On the other hand, the cloaca, which leads to the vent, is a highly functional duct. Not only does it take the excreted food products, it also has opening into it the ureters carrying urine from the kidneys, accommodates some of the secretions associated with reproductive activity, and in female birds it has the oviduct joining it. Here some water is recovered from the digesting mass before it is voided as faeces.

Birds have no bladder as such; urine becomes concentrated in the cloaca, where it precipitates uric acid, the white portion in droppings.

Respiration

The main respiratory effort for birds is in breathing out, not breathing in. Their very nature gives them a high oxygen requirement. As we have seen, some of their bones are pneumatic and this system extends to membranous air sacs, in addition to the lungs, through which air is pumped. These both lower the total specific gravity of the body and promote more efficient oxygenation of the blood.

Air enters via the nostrils and mouth, and a word of warning is in order here. Quite apart from the fact that de-beaking (beak trimming to prevent birds pecking each other) is too often an excuse for ignorant management, although it may sometimes be necessary, harsh de-beaking back to the nostril is positively injurious. Damaged nostrils mean an open beak and a clear passage for any harmful airborne organisms.

From the mouth the air travels down the *trachea*, a flexible tube of cartilagenous rings which divides into two primary *bronchi*. At this point is the *syrinx* or posterior larynx, which is the organ of voice production.

The avian lung has no 'bronchial tree'. It is firmly attached to the ribs. Each primary bronchus leading to a lung has three groups of secondary bronchi, many tertiary bronchi and numerous air capillaries.

Leading out of the lungs are the air sacs, some single, some paired. Sacs behind the lungs are the paired abdominal and posterior thoracic sacs. Air fills them from the primary bronchi. Sacs ahead of the lungs are the expiratory sacs, including the paired anterior thoracic and the single interclavicular and cervical sacs.

The senses

Eyesight is a highly developed sense in all birds. In *Physiology and Biochemistry of the Domestic Fowl*, P. E. King-Smith reports that a fowl's eyes together weigh almost as much as its brain, which is some indication of their importance. They occupy a much larger part of the head than outward appearance suggests.

Fast early growth is bred into broiler chickens such as this Ross 1

The position of the eyes gives the poultry species binocular vision— the ability to focus both eyes on a single object—but they can move them, and more frequently use them, independently. Neck mobility makes up for eye movements to a large extent.

Birds can see in colour, and domestic poultry have colour perception in a range similar to our own, though perhaps slightly favouring the red end of the spectrum.

Hearing is also well developed in birds, whose ears, holes in the side of the head, are protected by feathers. You have only to listen to a Mynah bird mimicking human speech to realize that they are able to define tiny inflections in sounds at least as well as we can. A good stockman will recognize this sensitivity; birds will respond favourably to a familiar voice or a whistle each day, but sudden sharp noises are obviously to be avoided.

The ear has three chambers, an outer ear, opening to the atmosphere, an air-filled middle ear and a fluid-filled inner ear which also incorporates the organs of balance. Orientation has to be a highly developed sense, of course, for any flying creature.

Endocrine glands

If you think of the central nervous system as the 'instant switch' of the body, giving immediate reactions to changes in the environment via the brain, the spinal cord, the peripheral nerves and the ganglia, it is easy to see the endocrine glands as the 'time switches', setting off slower but sometimes more profound reactions, to a precise schedule. They do this by releasing activating substances, the hormones, directly into the bloodstream and tissues.

Endocrine glands (the word, from Greek, means literally 'inward sifting') are ductless. Some of their purposes are clearly defined but others are more subtle, and the interrelationships can be highly complex.

The *pituitary gland* or *hypophysis*, which is a small structure at the base of the brain, has a controlling role over other endocrine glands through the variety of hormones it secretes. These include gonadotrophin, thyrotrophin and adrenotrophin influencing, respectively, the sex glands or gonads, the thyroid and the adrenal.

Growth rate, sexual development and the mechanics of egg production are all directed by secretions of this organ from its strategic position in close connection with that part of the brain known as the hypothalamus, and the optic centre.

Strange effects occur if the pituitary fails to produce gonadotrophin in male birds. This hormone acts on the testes, which themselves secrete testosterone, a hormone responsible for secondary male characteristics. If oestrogen, the female sex hormone, is injected into the male, it inhibits the pituitary from producing gonadotrophin. This process, known as caponization, is discussed in Chapter 14.

Pituitary activity in the growing bird is strongly influenced by light. An increasing light pattern—the sort of thing which happens naturally as days grow long from spring to summer—is detected by the bird's brain and passed on as a stimulus for the pituitary to trigger off increased secretions from those glands responsible for sexual maturity.

The important factor is the changing length of the light period from day to day, not its duration. If birds are growing during a period of increasing light they are likely to start laying at an early age, but the eggs will be small and immature. The hormone balance can be regulated, however, by adjusting the light pattern so that the birds experience a falling period of light each day. This keeps their reproductive ability in step with bodily growth, holds back the start of lay for from one to as much as four weeks, and ensures that eggs of a good size are available from the outset.

Light control is an important management skill dealt with at length in Chapter 6.

In the neck are the *thyroid glands*, one on either side of the trachea, which can vary considerably in size according to the strain of bird, the season—they tend to be larger in cold conditions—and iodine content. If they are removed at an early age, growth is stunted and a useless bird results.

Thyroid secretion has an important bearing on feathering, both the colouring and the moulting period being influenced by it.

Near the thyroid are the *parathyroid glands* which regulate the amount of calcium circulating in the blood. Their removal leads to rapid death.

A wide range of effects can be attributed to the *adrenal glands*, also known as the *suprarenal glands* from their position close to the kidneys and just behind the lungs.

Among their several secretions is adrenalin, familiar to most people as the hormone which sets the pulses racing after a fright. The metabolism of carbohydrates and proteins, the balance of sodium and potassium, and the regulation of blood pressure all come within the sphere of influence of the adrenal hormones.

The other leading function of the adrenal glands is in generating sex hormones which combine with those of the gonads to control sexual activity and the secondary sexual characteristics. In fact the adrenals are found close to the upper part of the sperm tract in males, while the left gland is usually embedded in the ovarian stalk of the female.

Within the pancreas, tiny groups of cells, the *islets of Langerhans* perform the very important job of secreting insulin, the hormone which determines the level of sugar in the blood and the development of glycogen to give energy to muscles.

Reproduction

Male

Two bean-shaped organs, the *testes*, are the originators of sperm in birds, and they stay well inside the body, near the head of the kidneys, never descending into an outer sac as in mammals.

They contain seminiferous tubules within which spermatozoa develop, and they also hold tissue which generates the male sex hormone testosterone. In the breeding season they enlarge dramatically from their resting phase.

Each testis gives rise to an *epididymis*, through which the spermatozoa pass to a long coiled duct, the *ductus deferens* which opens, at its lower end, into the cloaca.

With fowls and turkeys the opening to the cloaca is through a conical, erectile duct or papillus within the vent, but drakes and ganders, birds which commonly mate on water, have a developed penis which can enter the oviduct of the female.

Female

At the embryonic stage, while still in the egg, all birds have similar bodily characteristics, with two gonads, but by about the twelfth day of incubation their sexual destiny is already apparent, and in the female ovarian development starts. In the chicken only one ovary matures and this is almost invariably the left. The right remains rudimentary.

The physical capacity for egg production is tremendous, with huge numbers of ova available, of which only a relatively small proportion will eventually become complete eggs.

In the body of the reproductive female ova, at first microscopic in size, gradually accumulate yolk, the future nutrient for the embryo, and grow in a cluster around the ovary. Each is contained in a fine *vitelline membrane* and held in place by another thin membrane, the *theca folliculi*, which is well served with blood vessels.

When ovulation takes place the follicular membrane splits along the *stigma*, a line not crossed by any blood vessels, and the ovum is released into the body cavity. Occasionally, if the split is not quite precise, a blood vessel will rupture nevertheless, and a blood or meat spot results in the finished egg. Some birds appear to be more prone to this than others.

Now starts a journey of about twenty-four hours (in the hen, though similar periods apply in other poultry species). During this time the egg

Reproductive organs

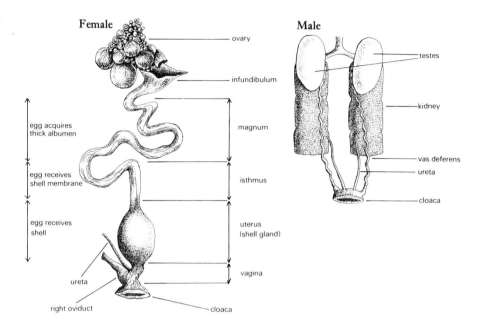

will increase in weight by about 70%, on its way through the oviduct, via the *infundibulum*, the *magnum*, the *isthmus*, the *shell gland* and the *vagina* to the cloaca from which it is laid. It is possible, just, for more than one egg to be laid within a twenty-four hour period but it remains very much the exception rather than the rule.

The infundibulum, corresponding to the fallopian tubes in mammals, is a widely flared funnel-like entrance to the oviduct which collects the ovum as it falls away from the ovary. At this point, if the bird has mated, the ovum will be met by spermatozoa and fertilized.

Birds have a sperm storage method peculiar to their system of reproduction. When mating, the female 'everts' the oviduct, so that from a position opening into the side of the cloaca it turns to face outwards at the vent opening and is brought into close conjunction with the male genitalia. Spermatozoa are deposited in the oviduct and make their way to tubular glands known as sperm-host glands at the upper end of the vagina, where they can survive for up to three weeks in the case of the fowl, though this period is variable. To fertilize each successive egg, sperm have to negotiate the full length of the oviduct and reach the infundibulum.

Whether or not it is fertilized, the ovum next moves into the longest section of the oviduct, appropriately called the magnum, where copious quantities of albumen are produced from glandular cells lining the walls. Here it first collects the dense inner albumen—that which stands up clearly, close to the yolk, when a new-laid egg is broken out. By this time it has also gained the rope-like chalazae fibres at each end which help to orientate the yolk in the centre of the egg.

The ovum is driven on by peristaltic squeezing movements of the oviduct, to the isthmus, where it receives the shell membrane, then to the shell gland, in which it remains for about twenty hours, increasing its coverage of albumen and putting on its calcium carbonate coat, which makes up about 10% of its finished weight.

The egg

Despite the 'bionic' fantasies of our age, it is still quite beyond our capabilities to manufacture something as simple as an egg, however much we spend on the attempt, because of course an egg in reality is anything but simple.

Consider the shell. In all species of poultry it is less than 1 mm in thickness (about 0.3 mm for chicken eggs) and of a relatively fragile material, yet thanks to its curvature it will withstand quite high stresses, while giving way to the hatching chick.

The outermost covering is the cuticle, a very thin layer of organic material which coats the shell and gives it its 'bloom'. One of its functions is to seal the shell against rapid moisture loss. It varies in thickness on different parts of the shell and between different strains of birds.

Beneath this lie the solid layers of the shell, the *palisade*, which is the outer and thicker portion, and the *mammillary* layer, which is usually between a third and a fifth of the total thickness. They are made up of innumerable columns of crystalline calcium carbonate or calcite. In the palisade layer these are packed shoulder to shoulder, but they open out into knob-like mammillae on the inner side of the shell.

Egg structure

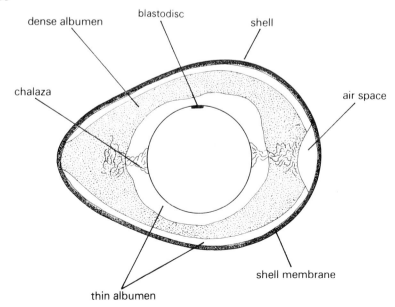

dense albumen · blastodisc · shell · chalaza · air space · thin albumen · shell membrane

Despite its solid appearance the shell has thousands of pores and is highly absorbent, a point of importance for both incubationists and those handling or storing eggs for sale.

Two membranes line the inside of the shell, the outer and thicker one being closely attached to the roughened mammillae. The thinner membrane adheres to the thicker except at the large end of the egg, where it separates to form a pocket, the air cell. As an egg ages, this cell increases in size and provides an indication of how old it is, because it can be seen against a strong light.

Four layers of 'white' or albumen surround the yolk, in alternate dense and liquid phases. Although the proportions vary substantially between eggs, the white accounts for nearly 60% of the total weight of an average egg, over 88% of this being water and around 10% protein.

Central to the egg is the yolk, histologically defined as an aqueous protein containing microscopic yolk globules and lipid or lipoprotein drops. What do you mean you will never eat one again! It is usually just under 50% water, about 30% lipid and 17% protein, with other compounds and inorganic elements.

Like the albumen, the yolk is stratified. It is formed in lighter and darker concentric layers, the result of the hen's reaction to cyclical changes in her twenty-four hour environment when the yolk is building up in the ovary. These colour differences are not always easy to see in normal yolks.

On the surface of the yolk is the all-important germinal disc which is known as the blastodisc in unfertilized eggs and the blastoderm when fertilized. From this develops the next generation.

3 A closer look at stock

Once you have decided whether you intend to keep layers, meat birds of whatever species or, say, bantams, the next step is to choose your stock. Which breed? Where from?

Until the 1950s it is fair to say that poultry-keeping was an occupation in which a time-traveller from the last century would not have felt unduly out of place. Progress had been made in a number of ways, but flocks in the main were still small, housed in barns or small portable sheds known as fold units, allowed on free range. Breeds and crosses went by old, universally recognized names like Leghorn and Rhode Island Red and a poultryman knew where he stood. Then came a change, and with almost revolutionary speed the old farm methods gave way to today's industry. New intensive methods requiring new, specially bred hybrids, took over.

Apart from those maintained by a few enthusiasts, the old breeds were swept away with quite surprising rapidity by the code-named newcomers. Yet in a way they were not entirely lost, because the hybrids owe their ancestry to these earlier lines.

The most obvious candidates for development were the fowls, both layers and meat strains, and turkeys. Selective breeding led quickly to fowls of light weight and low feed intake capable of laying high numbers of eggs in cages; and broilers having fast growth potential, again on low food intake, when reared on the floor in big deep-litter houses. From established lines of turkeys such as the Beltsville and American Mammoth Bronze came modern white-feathered birds with varying rates of growth and superior meat to bone ratio.

Now the professional can only turn to the hybrids produced by the major breeders if he is to remain competitive. Their performance is well documented both from the breeders' own tests and from independent random sample tests in various parts of the world. Recent figures for leading breeds can be found later in this chapter. The difficulty is that birds, being living creatures, tend to react according to their surroundings. On paper one breed can appear far more efficient than another, but under different management conditions it may be more susceptible to disease, more flighty, more finicky over rations—all factors that could tip the balance against it. Also, because the big inter-

Chickens like the Old English Game probably reached Britain before the Romans

national breeders do not stand still in their genetic work, today's 'top bird' has a habit of becoming tomorrow's 'has-been'.

Naturalists differ about the true origin of the modern fowl. One of the greatest, Charles Darwin, attributed all breeds of domesticated fowl to *Gallus Bankiva*, the wild jungle fowl found from Northern India to Burma and Malaya, but three other types of jungle fowl have been identified, *Gallus Stanleyii*, *Gallus Sonneratii* and *Gallus Furcatus*, and of these *Stanleyii* at least probably played a part in the modern fowl's ancestry. These wild birds weighed about 1.60 kg ($3\frac{1}{2}$ lb) and might lay twenty-five eggs in a season, but domestication started early, and a Chinese encyclopedia compiled around 1400 BC refers to chickens kept under artificial conditions.

Chickens looking quite similar to the modern Old English Game must have reached Britain in pre-Roman times, to judge from reports by Julius Caesar, but he noted that they were used by the Britons for diversion rather than the table. Ducks and geese would already have been native or regular visitants to the country, in the form of the Mallard and Greylag respectively. Later arrivals were turkeys and guinea-fowl, both of which were introduced in the sixteenth century.

The Romans themselves may have brought in the first truly epicurean bird to be seen in England. This would have been the forebear of the Dorking, a distinctive bird with five toes—most chicken have four— described in their texts. Dorking cockerels have been known to reach 6.35 kg (14 lb) in weight and this breed almost certainly provided the 'capons' which crop up frequently in records of historic banquets.

Over the years domestic strains grew up in several major world centres. Notably they came from the Mediterranean, which gave rise to the famous Leghorn, the Minorca and similar characteristically lively, prolific white-eggers; from Northern Europe which yielded such birds as the Sussex and Hamburgh, and of course from Central Asia itself. Then in the 1840s, from Northern China came the Cochins, also known as Shanghaes, large birds with feathered legs, to be followed by others from the same area, such as the Langshan. These birds introduced the factor for brown shelled eggs subsequently built into birds like the Maran, Welsummer and Rhode Island Red.

The Rhode Island Red was an American development, admitted to the American Standard of Perfection in 1904, which proved one of the most valuable assets to later breeders. It was a utility bird and it was this movement towards usefulness, started in the nineteenth century, and perhaps rather stronger in the US than in Europe, which led to the foundation stocks of the modern poultry industry. Hitherto, if breeds were developed at all, it was to win prizes at local shows by reaching certain 'standards' on, for example, the size and shape of comb, leg colour or intricacy of feather markings. No thought was given to economic factors.

The turning point, for English poultrymen at least, came with the introduction of a large, dual-purpose bird with black plumage by Orpington farmer William Cook. He named it after his town, the Black Orpington and it originated, he said, from the Langshan, Barred Plymouth Rocks from America, and Minorcas. He followed it with a white version in 1889, and the Buff Orpington in 1894, for which he crossed Cochin, Dorking and Hamburgh fowls. These Orpingtons were breeds 'manufactured' for farmyard use, and the Buff was capable of producing over 250 eggs in forty-eight weeks.

On the other side of the Atlantic the White Leghorn had already been taken up by the 1830s but was not destined to reach Britain until 1869, when it came over in the form of three prize-winning birds from the New York Show. By the 1880s another famous American breed, the Wyandotte, was being recognized for its egg-laying ability.

Black Orpington hen

It would be wrong to assume that nothing then happened to improve breeding performance until the advent of modern hybrids. There were many small-scale breeders and it was not uncommon to have upwards of twenty different lines on hand. These would be bred for show or to produce egg layers and meat birds for supply to local customers, but there was also a fresh awareness of the need to improve poultry performance. It extended from producer to official level.

In 1932 the Ministry of Agriculture in England introduced a voluntary system, the Accredited Breeders' Scheme, to encourage stock improvement. Breeders and hatcheries co-operating in the scheme were registered. This was merged in 1945 in a wider Poultry Stock Improvement Plan to guide what was fast becoming an organized industry through the difficult post-war years. Among other things, the PSIP included tests at the Official Progeny Testing Station, Wheaton Aston on the laying potential and profitability of pullets entered by different breeders.

Similar thinking in America led to the establishment of the National Poultry Improvement Plan in 1935. Administered by supervising agencies in each state and under the overall co-ordination of the USDA, this scheme set out to improve production and marketing factors, and put special emphasis on disease control at hatchery level.

Random sample tests, some official, like Wheaton Aston, others privately run, were popular at this time, breeders paying to get an independent assessment of their birds' laying abilities against others.

Testing involves keeping batches of birds from several competitors, sampled at random and therefore not necessarily the elite birds of a flock, under the same farm conditions. Very careful records of laying performance, mortality, food consumption, final carcass weight and profitability over a period which may take in rearing, are published.

As late as the 1960s small-scale commercial breeders were still fairly plentiful in England, some producing the old recognized breeds, White Leghorns, Rhode Island Reds and Light Sussex and their crosses, RIR × LS and WL × RIR. In the 500-day Commercial (Random Sample) Test of 1963–4 held by the National Poultry Tests, for example, thirty-four breeders took part with sixty-four entries, of which nine were

Ducks were in England before chickens, though these modern Aylesbury duckling are a little faster growing than their ancestors

crosses. But significantly, the remainder were hybrids or strains classified as 'experimental'.

Since then the competitive edge of the successful hybrids, and the increasing cost of running hybrid breeding programmes, has taken its toll and reduced the number of commercial breeders to a handful of big international companies. On the way it has also hit the number of independent tests, which had to adapt to the diminishing number of entries and the greater power of the remaining breeders, some of whom questioned the value of stock testing outside their own programmes.

Henry Wallace, in America, set the hybrid trend, and a new style in breed names, by developing the Hy-Line. The first bird to bear this name was derived from New Hampshire Red, Rhode Island Red and White Leghorn lines. He started with certain clearly defined targets in such factors as fertility, egg size and egg production required. First he mated each line pure and tested the progeny. Those coming closest to his targets were retained, to be mated brother to sister, the rest were disposed of or 'culled'. This procedure was repeated for several generations, the best being kept and inbred each time. Finally the offspring of each line were crossed to produce the new commercial line.

Hybridization of this type is described in more detail in Chapter 4, but multiple crossing, expensive though it is in terms of research and wastage, has led to virtually all the modern industry's commercial layers and broilers and a proportion of its ducks and turkeys.

Physical selection

When buying stock in the quantities that the professional egg or poultry meat producer buys them, it is obviously impractical to inspect them in detail beforehand. Selection, therefore, boils down to knowing the record of the strain, and the supplier. This may be a hatchery or a rearer, depending on whether the poultryman is taking chicks—which he would obviously do with broilers for instance—or grown birds.

In the first instance the professional will turn to the breeder's prospectus, supplemented if possible by independent test reports, for his facts. Some of the 'vital statistics' of a number of leading commercial breeds are set out in the tables on pp. 34–38, but again it must be emphasized that figures can only offer a general guide, subject to conditions. Given top management a 'bad' bird on paper may do better than a 'good' one less well handled.

The experienced poultryman knows his market and this will influence his choice to some extent. If he is an egg producer he will know whether his customers prefer brown or white eggs and whether they will readily pay more for large eggs. This will help him to determine whether he is going to have white-eggers, which are generally lightweight birds, or layers of tinted or brown eggs, derived from medium weight strains; but that brings in other considerations. White eggs are more easily candled for cracks and internal quality than brown. White birds tend to eat less and produce more eggs than brown, but they are also inclined to be more nervous and temperamental. Brown birds tend to produce more large eggs and withstand stresses a little better. These are the 'tendencies' but it is a fluid situation, and in recent years brown birds have come much closer to duplicating the performances of white.

Between competing strains of white-eggers or brown-eggers, of course, the choice is even narrower. Here, apart from a chat to visiting feed representatives or local advisers, which can be instructive, random sample test results published in poultry magazines as they are issued, can help. In America results are aggregated from a number of tests in various parts of the country, in the USDA summary.

So much for the well-documented hybrids, but what if you deliberately choose to buy a traditional breed with good—but slow-gaining—meat qualities, or a pure-bred layer? In buying a limited number of stock you want to be sure that you are getting value for money. This is the sort of purchase which can call for personal inspection.

At the outset, the variety of breeds can be a bit breathtaking, but they can be rationalized into categories. Those listed in the following table are just a few from the many show varieties, which can be thought of as utility birds, and details can be obtained from the breed clubs or individual breeders, who can often be met at shows. It is probably best, other things being equal, to go for breeds closest to your own neighbourhood. In addition to the breeds listed, you also have the rest of the exhibition categories, including the bantam classes to choose from. Bantams are generally miniatures of the full-size birds, showing similar breed characteristics.

General purpose (large bodied birds but with laying ability)		Layers	
		Ancona	White eggs
		Hamburgh	,,　,,
Faverolle	Tinted eggs	Leghorn	,,　,,
Houdan	White eggs	Minorca	,,　,,
Maran	Brown eggs	Welsummer	Brown eggs
New Hampshire Red	Tinted		
Orpington	Brown	Table breeds	
Plymouth Rock	Tinted	Aseel	Tinted eggs
Rhode Island Red	Brown	Dorking	,,
Sussex	Tinted	Indian Game	,,　,,
Wyandotte	,,	Jersey Giant	Brown eggs

Methods of purchase

Several methods of purchase are open to the poultryman. Hatching eggs are obtainable for all species of poultry, at reasonable prices. An incubator is needed, or a broody hen, and there is a gamble both on the fertility of the eggs and the sex of the chicks, but there is great satisfaction in a successful hatch.

Day-old chicks are probably the most popular way for the amateur to start a flock. Again reasonably cheap from hatcheries, pet shops or markets, the young birds look attractive and are easy to carry. They are something of a gamble, however. Vulnerable to chilling and infections, they need constant care in the early stages, and the inexperienced, buying from an unknown source, can find they have bought all cockerels.

The next alternative for the egg producer is to buy birds in at point of lay, around eighteen to twenty-two weeks of age. They cost more, naturally, because they have been brought through the crucial early stages, but if they are properly reared commercial layers they will have been vaccinated and given specialist attention, and are ready to begin their productive life, without, we hope, any further worries or delays.

Some people advocate buying at intermediate stages in growth, at six weeks for example, but there is little advantage in this. It means taking on immature stock, breaking their routine and, more crucial, risking a change in their lighting pattern. Even with turkey poults, ducklings and goslings it is advisable to buy birds as near as possible to one day old and 'start as you mean to go on'.

Occasionally poultrymen will buy hens after a season in lay. You get fewer, but larger, eggs after the first moult, and they should cost less than pullets. These are inducements, but beware of purchasing grown birds unless you know their history. Plenty of beginners have bought old hens, sometimes three or four years old and long past laying prime, believing them to be year-old or perhaps even point of lay birds.

Learn to look at birds. An inspection of the legs and feet is a useful first check on the age and condition of a bird. Leg scales become more horny and lose their lustre as age progresses. Feet and claws harden and become more grimed.

Feathers form in a definite order, and adult primaries may not be complete before the laying stage is reached, but strains vary in their speed of feathering, and this alone would be an unreliable method of telling, for instance, if you were looking at a point of lay pullet. Nevertheless, good feathering is another guide to general condition.

Handling birds tells the stockman a number of things. A young pullet feels light and compact. The rear end of the breast-bone is pliable. This is because the sternum is made up of two types of bone, the front end hard but the rear end cartilagenous and almost translucent in young birds. It is something to be noticed when carving a broiler, which is a bird killed at around eight weeks before the tail end of the sternum has ossified, which it does gradually until, approaching point of lay, just the rear tip is still soft.

It should also be possible, when holding a hen, to gently feel the spread of the pelvic bones. In the layer with good potential they move apart and in full lay may allow space for three to four fingers. Space between the end of the sternum and the pelvic bones is also an important factor, with a bearing on internal volume, and therefore laying capacity, of the hen.

In a hen in full lay, four fingers can be placed between the pelvic bones and the end of the breast bone. If there is only room for two fingers, the bird is not laying.

Wing feathers
1—shoulder butt; 2 and 5—bow or coverts; 3—bar; 4—secondaries; 6—axial feather; 7—primaries

Dotted line indicates approximate cutting area when wing clipping. More drastic, but more permanent, is removal of first wing joint when birds are under ten days old.

1   2   3   4   5   6   7

As a general rule, whatever age or species of stock you are buying, look for lively birds with a bit of 'pride' about them, carrying their heads well and taking an active interest in their surroundings. Observe the eyes. They should be as wide, bright and prominent as you would expect for the species. Beware of dull, sunken eyes appearing to be lower or farther back than average for the breed. And remember that evidence of some nutritional deficiencies and diseases shows rapidly in eye conditions.

Features of some leading hybrid layers

Name of company Code of breed	Feather colour	Hen-housed average (52 wks) (Target)	Body-weight		Food consumption		Feed per doz. eggs
			20 wks	76 wks	0–20 wks	20–72 wks	
White egg strains							
BABCOCK B-300V	white	270+	(lb)2.5–2.75 (kg)1.1–1.2	3.9–4.3 1.8–1.9	16.5–17.5 7.5–7.9	88.7–93.3 40.3–42.3	4.00 1.82
DEKALB XL-Link	,,	255–280	2.75–3.00 1.24–1.36	3.96–4.18 1.8–1.9	15.6–16.7 7.1–7.6	88.00–97 39.95–44	3.96–4.4 1.8–2.0
EURIBRID Hisex White	,,	288	2.90 1.31	3.90 1.77	16.3 7.4	92.1 41.8	4.18 1.89
H & N Nick Chick	,,	272+	2.6–2.9 1.2–1.3	3.96–4.3 1.79–1.95	16.5–17.5 7.5–7.9	86.45–95.5 39.2–43.4	3.8–4.1 1.7–1.9
HUBBARD White Leghorn	,,	260+	3.00 1.36	4.4 2.0	14–16 6.4–7.3	89–95 40.4–43.1	4.0–4.1 1.8–1.86
ROSS Ross White	,,	275	2.9[1] 1.31	4.4 2.0	14.3[2] 6.49	88–98[3] 40–44.5	4.2 1.9
SHAVER S 288	,,	280+	2.7–3.0 1.2–1.4	4.0–4.3 1.8–1.9	14–16 6.3–7.3	92.1 41.8	3.8–4.1 1.7–1.9
THORNBER C2	,,	284	3.25 1.47	4.5 2.04	14.8[2] 6.7	88.72 40.28	3.8–4.1 1.7–1.9
Tinted egg strains ROSS Ross Tint (Sykes)	white flecked brown	270	3.2[1] 1.45	5.1 2.31	15[2] 6.8	92.5–103.5 41.9–46.9	4.4 2.0

[1] At 18 weeks   [2] 0–18 weeks   [3] 18–72 weeks

Hen-housed average is based on the total eggs produced in a 52-week laying period, divided by the number of pullets housed at point of lay. The figures quoted in this table are based on breeders' estimates, but can vary widely, of course, under differing management conditions.

Name of company Code of breed	Feather colour	Hen-housed average (52 wks) (Target)	Body-weight		Food consumption		Feed per doz. eggs
			20 wks	76 wks	0.20 wks	20–72 wks	
SHAVER Starcross 444	white	250–270	3.0–3.3 1.4–1.5	4.3–4.7 1.9–2.1	15–17 6.8–7.7	91–102.4 41.3–46.5	4.0–4.3 1.8–1.9

Brown egg strains

Name of company Code of breed	Feather colour	Hen-housed average (52 wks) (Target)	Body-weight		Food consumption		Feed per doz. eggs
			20 wks	76 wks	0.20 wks	20–72 wks	
ARBOR ACRES Brown	black and gold	263+	3.7 1.68	5.0–5.4 2.3–2.5	17 7.7	90 40.86	4.4–4.6 2.0–2.09
BABCOCK B380	light to mid- brown	260–275	3.5–3.75 1.6–1.7	4.98 2.26	17.5–18.5 7.9–8.4	93.3–97.8 42.3–44.4	4.23 1.9
DEKALB Amber Link	white with brown	270+	3.5–3.75 1.6–1.7	5.25 2.38	20 9.08	100 45.4	4.36–4.75 1.98–2.15
EURIBRID Hisex Brown	white with brown	272	3.59 1.62	4.9 2.2	16.8 7.6	98.5 44.72	4.38 1.98
HUBBARD Golden Comet	light brown over white	260+	3.6 1.63	4.73–5.1 2.15–2.3	15–16[2] 6.8–7.3	91 41.3	2.75–2.9 1.24–1.3
ROSS Ross Brown	brown	270	3.2[1] 1.45	4.4–4.6 2.0–2.08	15.4[2] 6.8	93.5–103.5[3] 42.4–45.8	4.4 2.0
Ranger	brown	245+	3.2[1] 1.45	4.4–4.6 2.0–2.08	15[2] 6.8	92.5–101[3] 41.9–45.8	4.6 2.08
SHAVER Starcross 579	light brown white flecks	265+	3.4–3.6 1.5–1.6	5.0–5.4 2.3–2.5	16–18 7.26–8.2	91–100.1 41.3–45.4	4.4–4.6 2.0–2.08
Starcross 585	brown	265+	3.5–3.9 1.6–1.8	5.3–5.7 2.4–2.6	16–18 7.26–8.2	4.1–4.5[4] 116–128	4.4–4.6 2.0–2.08
TETRA SL	reddish brown over white	255–270	3.6 1.6	5.0–5.25 2.27–2.38	16.6 7.53	90–94 40.86–42.7	4.2–4.45 1.9–2.02
THORNBER CT 404	brown	260–275	3.7 1.67	4.75–5.0 2.15–2.3	16[2] 7.26	95.5 43.36	
WARREN (ISA) Sex-Sal Link	reddish brown over white	280+	3.5–4.65 1.58–2.1	5.02–5.51 2.28–2.5	20 9.08	100 45.3	4.5 2.04

[1] At 18 weeks   [2] 0–18 weeks   [3] 18–72 weeks
[4] Oz consumed per hen per day during lay

Performance of some commercial broiler strains

Name of company Code of breed			Age (days)			
			42	56	63	70
COBB						
500	males	liveweight (lb)	3.3	4.9	5.6	6.1
		(kg)	1.50	2.22	2.54	2.77
	females	,, (lb)	2.7	3.9	4.5	4.7
		(kg)	1.22	1.77	2.04	2.13
Food conversion ratio (as hatched)			1.7–1.9	1.9–2.3	2.1–2.4	2.4–2.6
EURIBRID						
Hybro	males	liveweight (lb)	3.16	4.65	5.44	6.23
		(kg)	1.43	2.11	2.47	2.83
	females	,, (lb)	2.75	3.90	4.49	5.06
		(kg)	1.25	1.77	2.04	2.30
FCR (as hatched)			1.89	2.10	2.21	2.32
H & N						
Meat Nick	males	liveweight (lb)	3.41–3.51	5.03–5.17	5.75–5.95	
		(kg)	1.55–1.59	2.28–2.35	2.61–2.70	
	females	,, (lb)	2.81–2.91	4.03–4.17	4.64–4.84	
		(kg)	1.27–1.32	1.83–1.89	2.11–2.20	
FCR (as hatched)			1.75–1.90	1.90–2.10	2.10–2.30	
HUBBARD						
Hubbard	males	liveweight (lb)	3.25	4.85	5.60	
		(kg)	1.47	2.20	2.54	
	females	,, (lb)	2.80	4.00	4.60	
		(kg)	1.27	1.81	2.09	
FCR (as hatched)			1.72	2.11	2.46	
MARSHALL						
'M'	males	liveweight (lb)	3.65	5.30	6.05	6.80
		(kg)	1.65	2.40	2.75	3.09
	females	,, (lb)	3.05	4.30	4.85	5.40
		(kg)	1.38	1.95	2.20	2.45
FCR (as hatched)			1.93	2.13	2.21	2.29
ROSS						
Ross 1	males	liveweight (lb)	3.3	5.1	6.0	6.9
		(kg)	1.5	2.31	2.72	3.13
	females	,, (lb)	2.7	4.1	4.8	5.35
		(kg)	1.22	1.86	2.18	2.43
FCR (as hatched)			1.81	2.05	2.21	2.35
SHAVER						
Starbro	males	liveweight (lb)	3.7	5.3	5.9	6.8
		(kg)	1.7	2.4	2.7	3.1
	females	,, (lb)	3.0	4.2	4.8	5.5
		(kg)	1.4	1.9	2.2	2.5
FCR (as hatched)			1.8–1.9	2.0–2.1	2.2–2.3	2.3–2.4

Turkeys

Name of Company Code of breed			12	16	Age in weeks (food conversion in brackets) 18	20	22	24
ARNEWOOD								
Double AA	male l/wt	(lb)	14.4(1.9)		23.8(2.39)			32.7(3.14)
		(kg)	6.5		10.8			14.9
	female	(lb)	10.97(2.1)		16.1(2.74)			21.7(3.45)
		(kg)	4.85		7.35			9.87
Treble CCC	male l/wt	(lb)	11.0(1.94)		18.6(2.7)			26.1(3.2)
		(kg)	5.0		8.5			11.9
	female	(lb)	7.6(2.34)		12.4(2.9)			17.0(3.52)
		(kg)	3.48		5.63			7.74
Mini de luxe	male l/wt	(lb)	9.2(2.15)		16.6(2.6)			19.0(3.55)
		(kg)	4.18		7.56			8.63
	female	(lb)	7.7(2.37)		10.25(3.1)			11.75(3.8)
		(kg)	3.5		4.77			5.34

Name of company Code of breed			Age in weeks (food conversion in brackets)					
			12	16	18	20	22	24
ATTLEBOROUGH POULTRY FARMS Broadacre White	male l/wt	(lb)	10.7(2.14)		17.8(2.59)			25.6(2.94)
		(kg)	4.86		8.08			11.62
	female	(lb)	8.1(2.17)		13.1(2.74)			
		(kg)	3.6		5.94			
Wrolstad White	male l/wt	(lb)	10.0(2.08)		16.6(2.58)			23.4(2.99)
		(kg)	4.54		7.53			10.62
	female	(lb)	7.6(2.19)		11.46(3.00)			
		(kg)	3.45		5.20			
BRITISH UNITED TURKEYS Triple 6	male l/wt	(lb)	13.52(2.01)		23.8(2.55)			32.51(3.34)
		(kg)	6.14		10.80			14.76
	female	(lb)	10.4(2.27)		16.64(2.89)			
		(kg)	4.72		7.55			
Triple 5	male l/wt	(lb)	10.85(2.14)	15.9(2.52)				20.3(2.94)
		(kg)	4.92	7.21				9.21
	female	(lb)	8.00(2.36)	11.2(2.84)				13.2(3.45)
		(kg)	3.63	5.08				5.99
HAMMOND Heavy	male l/wt	(lb)	11.6(2.00)		20.6(2.5)			29.8(2.93)
		(kg)	5.26		9.35			13.52
	female	(lb)	8.8(2.1)		15.7(2.61)			
		(kg)	3.99		7.13			
Medium	male l/wt	(lb)	10.8(2.04)		18.9(2.58)			27.0(3.11)
		(kg)	4.90		8.58			12.26
	female	(lb)	8.3(2.16)		14.3(2.77)			
		(kg)	3.77		6.49			
INDICO Maxi	male l/wt	(lb)	12.3(2.0)	18.47(2.5)	21.4(2.7)	27.4(3.1)		
		(kg)	5.59	8.38	9.71	12.43		
	female	(lb)	9.7(2.2)	13.7(2.6)	14.9(2.9)			
		(kg)	4.39	6.2	6.75			
Midi	male l/wt	(lb)	10.4(2.0)	16.3(2.4)		21.2(2.7)		
		(kg)	4.73	7.4		9.61		
	female	(lb)	8.5(2.0)	11.9(2.5)				
		(kg)	3.88	5.42				
Mini	male l/wt	(lb)	9.6(2.0)	14.6(2.4)				
		(kg)	4.38	6.62				
	female	(lb)	7.3(2.2)	8.8(2.4)				
NICHOLAS Large White	male l/wt	(lb)		19.6(2.3)		24.9(2.7)		30.8(3.25)
		(kg)		8.9		11.3		14.0
	female	(lb)	9.9(2.2)	14.3(2.5)		17.6(3.1)		
		(kg)	4.5	6.5		8.0		
ROSS Super Midi	male l/wt	(lb)	12.46(2.07)		20.83(2.6)			27.67(3.17)
		(kg)	5.65		9.45			12.55
Midi	male l/wt	(lb)	11.13(1.98)		18.4(2.52)			25.47(3.06)
		(kg)	5.05		8.35			11.55
	female	(lb)	8.69(2.24)		14.24(2.9)			
		(kg)	3.94		3.94(6.46)			
SUN VALLEY S.V.II	male l/wt	(lb)			16.15(2.28)[1]	22.51(2.82)		
		(kg)			7.34	10.23		
	female	(lb)			11.55(2.51)[1]			
		(kg)			5.24			

[1] At 15 weeks.

Ducks

Name of company Code of breed			Age in days (food conversion in brackets)	
			47	56
CHERRY VALLEY Commercial strain White feather				
L3	liveweight	(lb)	6.8(2.81)	
		(kg)	3.09	
L2	,,	(lb)	6.5(2.82)	
		(kg)	2.97	
M1	,,	(lb)	5.9(2.85)	
		(kg)	2.68	
S2	,,	(lb)	5.0(2.89)	
		(kg)	2.27	
S1	,,	(lb)	4.2(2.92)	
		(kg)	1.92	
Coloured feather				
CL2		(lb)	4.89(2.94)	
		(kg)	2.22	
CM1		(lb)	4.17(2.92)	
		(kg)	1.89	
CS3		(lb)	2.84(2.98)	
		(kg)	1.29	
CS4		(lb)	4.17(2.92)	
		(kg)	1.89	
HEJGAARD				
Heavy	liveweight	(lb)		7–7.3
		(kg)		3.2–3.3
Medium	,,	(lb)		6.5–6.7
		(kg)		2.9–3.1

4 Breeding for fun

For true poultry fanciers there are few thrills to beat taking a 'first' at a poultry show for a bird they have bred and reared themselves. It is a recognition of all the time spent in observing, recording, selecting and ultimately producing something more close to perfection than any other entry in its class. And it demonstrates the value of the stock to other breeders, including 'industrial' geneticists, who are sometimes glad to call on new blood in making up their hybrids.

Show breeding for fun is perfectly feasible as a back garden operation. It does not require large numbers of birds, but it does demand time, a basic understanding of genetics and enthusiasm.

There are several ways of entering this side of poultry-keeping. Quite often it is a family interest, with experience and good stock being passed on to the next generation. Sometimes people are 'hooked' at shows when they see the colour and variety of the poultry breeds for the first time and perhaps think they could do better in one class or another. Shows, incidentally, usually supply a strong incentive for the young, with special juvenile sections. If you can visit a friendly breeder in your locality he may furnish you with stock and worthwhile advice.

Failing all these routes, your most obvious first move is to contact your national poultry club. In the UK this is the Poultry Club of Great Britain and you should write to the Secretary at Virginia Cottage, 6 Cambridge Road, Walton-on-Thames, Surrey. In the US there are the American Poultry Association, P.O. Box 70, Cushing, Ok. 74023 and the American Bantam Association, P.O. Box 610, N. Amherst, Mass. 01059.

These clubs provide show timetables and the all-important information on specialist breed clubs and how to reach them. The national clubs are long established and have done much to co-ordinate the interests of the many specialized breed clubs. A standard is issued by each club, setting out all the distinctive features for its particular breed, as a guide to breeders and judges alike. Hence the origin of 'standard' breeds.

By definition, a pure breed is a group having certain clear characteristics which are inherited from generation to generation and distinguish it from any other breed. These are its racial characteristics. Within its type, however, there may be differences in feather colour or markings

Silver spangled Hamburgh cockerel. The rose comb is characteristic of all members of the breed, but colour varies widely

which make up separate varieties. A good example is the Hamburgh, a light breed of fowl with good body proportions. In ideal form it should have large wings, a long sweeping tail, a short, well curved beak, neck and legs of medium length and four toes. Its most obvious distinguishing feature is a rose comb, appearing in both male and female. These are all breed characteristics, shared by every Hamburgh true to its breed. But there are Black, Pencilled, Gold Spangled and Silver Spangled varieties (pencilling and spangling being descriptions of feather marking) which look widely different from each other, though they are all Hamburghs.

Breeders mate and select birds to come as close as possible to purity of standards for the breed and it is up to the judges, awarding points, to determine how close they get. The scale of points differs for each variety but always adds up to a possible 100. Head features of the male Black Hamburgh, for instance, comb, face and ear-lobes, are worth a maximum of forty-five points, its colour twenty-five. The Pencilled variety, on the other hand, gains only twenty-five for head features, but a total potential of sixty-five for markings.

Standards are applied in a similar way to all recognized poultry breeds including ducks, geese and turkeys. To see them illustrated and explained you should look at *British Poultry Standards*.

Occasionally new breeds and varieties are still originated. To have them accepted as Standards, living examples of at least two generations must be supplied to the Poultry Club for consideration by its experts, with a sworn declaration that the breed produces true; in other words, that the progeny duplicate the characteristics of the breed. The only exceptions to this standardization procedure are breeds which have been accepted as Standards by poultry clubs of other countries.

Mendelism

Gregor Mendel, abbot of an Austrian monastery from 1860 to his death in 1884, was the rather unexpected founder of the principles of heredity on which modern genetics is based. His *Law of Independent Segregation of Factors* has had its critics, but it still stands as a fundamental guide to breeders.

What Mendel established, by experiments with plants, applies with equal force to birds and indeed all living organisms which have characteristics laid down by their progenitors. This is the pattern by which parents pass individual features to their offspring, and the way in which certain of these features from one parent will give way to more dominant features from the other.

With the fusion of a sperm cell or male 'gamete' and an ovum or female 'gamete', a fertilized cell known as a 'zygote' is formed. Each gamete brings with it a package of chromosomes, representing a sampling of half of the parent's inheritance. These chromosomes are microscopic strings of genes carrying blueprints for the new individual. The chromosomes have specific roles in controlling future development, and those from one parent pair off with those from the other, creating in the zygote a nucleus of paired chromosomes (and therefore paired genes). Multiplication is marked by the chromosomes splitting lengthwise, the two halves drawing to opposite sides of the zygote, which then divides into two cells, each with its own nucleus of carbon-copy chromosomes. Then the process continues, two cells becoming four and so on, but always carrying the same mixture of genetic information from each parent.

Mendel showed how this information was interpreted. With his plants he demonstrated that certain characteristics, such as tallness, were dominant. A tall plant crossed with a short plant resulted in tall offspring. But if breeding continued, using the tall offspring, a mixture of tall and short in the ratio of 3:1 appeared in the new generation. Clearly what one could see was not the whole story, otherwise breeding two talls would result only in tall descendants.

Large single comb

Scientists call the body one can see the 'phenotype' and the invisible genetic potential of a parent its 'genotype'. The fact that an individual is tall does not necessarily mean that their genotype is for tallness. In their cell structure they inherited a pair of chromosomes, one from each parent, containing genes to decide their stature. These genes may both be for tallness, in which case the individual is pure-bred (*homozygous*) for tallness and will pass on that characteristic to the next generation. But suppose the individual receives a gene for shortness from one parent? Because it has a dominant gene for tallness it will be tall, but it will pass on a mixed inheritance of tall and short characteristics in the form of a 'recessive' gene to the next generation. It is said to be impure or *heterozygous*.

In the fowl it is quite easy to pick out some dominant bodily characteristics. The tendency to rose combs is dominant to single combs, for instance, and white skin to yellow. But can you tell, just by looking at it, whether a bird will breed pure or not? Paradoxically, you can if it shows recessive features. A bird with a single comb can only be homozygous, having matching genes to produce the single comb, because if it had the dominant rose comb gene, it would have to show it. More difficult is the bird which shows a dominant feature since, as we have seen, that is no guarantee of its genetic makeup.

Rose comb

Testing for purity

Given a reliable record of a bird's ancestry, the breeder can be reasonably confident of how it is going to reproduce in terms of simple features. Quite often, however, birds are acquired on a casual basis by the enthusiast, perhaps as a batch in which two or three show promise. Then it might be thought necessary to test for genetic purity.

This form of testing relies on mating the unknown to the known and studying the progeny. We know a bird with recessive characteristics must be homozygous. The bird under test will, of course, show its domi-

Rose comb with leader

nant factor. If it is also pure and the two are mated, all the offspring will carry the dominant characteristic. A heterozygous bird, on the other hand, mated to the pure recessive, will produce 50% of chicks with a recessive factor and 50% with a dominant factor.

While it is useful to know the purity of stock, a recessive gene is not always an undesirable factor. Sperm from a homozygous rose-combed male, for instance, appears to have a survival time in the oviduct of about half that of sperm from single-combed or heterozygous rose-combed males. Some breeders are prepared to accept the recessive factor in these cases, in return for superior fertility.

Sex linkage

Sex determination is a further complication in the genetic picture, for this is the one point where the pairing of chromosomes so far described is modified. To understand this it is necessary to look again at the way in which the chromosome pairs are established.

Gametes, the sperm and ovum, each carry a half-set of chromosomes from the sire and dam respectively. When fertilization takes place and the zygote is formed, the chromosomes come together to make the full set for the new individual, but its sex is laid down by the presence or absence of a chromosome from the dam. If she does pass on a chromosome—often referred to by geneticists as the X chromosome—it will pair with the X chromosome from the sire and resulting offspring will be male, having a full complement of paired chromosomes. If, however, the X chromosome from the dam is missing, or possibly replaced by a dissimilar Y chromosome which some scientists claim to have identified, the offspring will be female. In practice, since every other ova carries an X chromosome, the balance between numbers of males and females is very close.

This Sebright of the gold variety, and the birds in the following picture contrast the difference between the two varieties

A number of genes are carried on the X chromosome, and because they are thus 'sex-linked', their effects show up irrespective of the dominant/recessive principle which applies to genes on other chromosomes. Before the genetic background was understood, poultrymen had already noted that in matings of certain lines, males and females had different feather colour which helped to pick out pullets from cockerels even at day-old.

The reason behind this, it is now known, is the operation of genes in the X chromosome which affect feather colour. Fowls, in general, come into one of two categories, either 'gold' or 'silver' colouration. This is a genetic definition that takes into account the many variations in between. A typical gold breed is the Rhode Island Red and a typical silver the Light Sussex.

If a Rhode Island Red cockerel with two recessive genes for gold colouring is mated to a Light Sussex hen with a dominant silver gene, it might be thought that the dominant gene would come through and make the offspring silver. Indeed it does, but for the males only. They

Sebrights of the silver variety

alone receive the dominant gene from their mother and so appear as yellow chicks. The females receive no gene from her, so the recessive gold gene from the father dictates their colouring, and they hatch brownish, which makes them easy to distinguish from their yellow brothers.

A number of other breeds have similar characteristics, such as:

Gold	**Silver**
Brown Leghorn	Silver Dorking
Buff Orpington	Duckwing Leghorn
Gold Spangled Hamburgh	Ancona
Golden Wyandotte	Salmon Faverolle
Indian Game	Silver Laced Wyandotte
Golden Campine	Columbian Wyandotte
Red Sussex	Silver Spangled Hamburgh

The silver-gold combination works only for a mating of a gold cockerel to a silver hen. A reciprocal mating, with a Light Sussex cockerel carrying dominant silver genes and a Rhode Island Red hen carrying a recessive gold gene, for example, results in all offspring being silver, while a heterozygous silver male and a recessive gold female leads to a mixture of silver and gold chicks, both male and female.

From what has been said, it would be wrong to assume that colour is always determined by sex-linked genes. 'Autosomal' genes—those other than the genes on the sex chromosomes—can also play a part, and sometimes quite a powerful one. The dominant white of the White Leghorn, for instance, results from an autosomal gene, and can override the effects of the sex-linked genes for colour.

Barring, the characteristic light and dark 'cuckoo' feather pattern found in Barred Plymouth Rocks, Marans, Scots Greys and North Holland Blues, is influenced by another sex-linked gene. Some other breeds, such as the Campine and Pencilled Hamburgh, also have barring, but in their case it is determined by an autosomal gene, and therefore has no value as a sexing method.

Here again, a hen of a barred variety, if mated with an unbarred cockerel such as a Black Minorca, passes her barring to sons only, the daughters all being black like the father. In the chicks, the barring character appears as a light patch on the head, all the chicks with this naturally being males.

Providing the crosses produce black chicks, this is an effective means of identifying the sexes. So the hens of Barred Plymouth Rocks, North Holland Blues and Marans all produce black chicks if mated with Light Sussex, Rhode Island Red or Brown Leghorn males; but the Legbar hen, which also has barred plumage, produces brown chicks and confused markings when mated in the same way with, say, a Rhode Island Red.

Leg colour and speed of feathering are two more features subject to sex-linkage. Crossing a white La Bresse male with a White Wyandotte female will give you male chicks with yellow shanks and females with blue, but it is not always easy to pick this out at one day old.

The feathering difference appears where rapid feathering breeds, usually lightweights such as the Leghorn, are crossed with breeds having slower feather development, like the Rhode Island Red. Male chicks from a rapid feathering sire and a slow feathering dam are themselves slow feathering and examination of a wing shows undeveloped quills all of approximately the same length. Pullet chicks, on the other hand, will already have primaries extending beyond the other wing feathers.

Feather sexing can be applied commercially, although most hatcheries employ professional sexers to separate pullets from cockerels. The skill of vent-sexing was developed in Japan, and sexing teams are still, more often than not, Japanese. By opening or everting the vent it is possible to recognize the male or female configuration in the cloaca. Both sexes have a tiny lump or prominence set in the folds of tissue near the entrance to the vent, but with the male it is usually slightly larger and more clearcut than with the female. An experienced sexer will make the correct decision with no more than 5% error, at speeds up to 1,000 chicks an hour, but it calls for dedicated training and great concentration, so good sexers are in demand. As a result vent-sexing is an expense to hatcheries and the teams, which travel from one to another, do not operate everywhere. Some of the major breeders have taken this into account by developing slow feathering strains for parts of the world which may lack professional sexers. The Babcock B300F was bred along these lines, for example.

Feather sexing, showing different rate of quill development; female (*top*) more advanced than male (*bottom*)

Autosexing

Recognition of the various sex-linked factors early in the century led to the development of trade in sex-linked crosses giving pullets for laying and cockerels for table, but the technique created the need to keep two separate breeds, and the sex-linked hens could not, in the usual way, be used for further breeding.

Breeders of the time wanted a bird which would breed pure while still showing sex-distinction at hatching. It came as a result of work at Cambridge University by Professor R. C. Punnet and M. S. Pease. Punnet, one of the pioneers of poultry genetics, was a partner of William Bateson, who was researching the subject in 1898, even before Mendel's paper on heredity was published in English in 1900.

Together, in the 1920s, Punnet and Pease were studying autosomal barring of birds like the Campine and wondered what would happen if this was crossed with birds carrying sex-linked barring. They tried it with the Barred Rock and Gold Campine, which has a mottled brown down, recessive to black. The first cross chicks hatched with black down and a light head patch. In the second generation, chicks had brown colouring. Those which exhibited a light patch were retained as combining the dominant sex-linked barring and the recessive Campine barring. They proved heterozygous for the sex-linked barring. When mated together to obtain homozygous sex-linked barring, these birds produced male chicks in which the light patch had spread over the neck and back, while the females showed dark, mottled down. This was the first autosexing breed, the Gold Cambar, and it was put on show at the World's Poultry Congress in 1930 at the Crystal Palace.

Breeders soon recognized that the method could be applied to other lines, some with better egg production potential than the Campine and new autosexing breeds were introduced, each with the suffix -bar after it. In this way the Legbar (Leghorn cross), Dorbar (Dorking cross) and Welbar (Welsummer cross) came about, with a number of others now accepted as standard breeds.

Up to the 1950s a bright future was predicted for autosexed birds, but the improving efficiency of vent sexing, applied to the new hybrids then coming in, drove them off the commercial stage before they had gained a footing on it. Nevertheless they still have their devotees among small-scale breeders, often exceeding the performance of the original pure breeds and giving clear sex distinction in the chicks regardless of which way they are crossed.

Sex-linkage is not solely limited to fowls. Ducks, geese and turkeys all have genes which in some strains influence early colouring, but two things make these factors less significant than in the fowl. One is the strong commercial movement to white turkeys and ducks. The other is the relative ease with which these species can be vent sexed.

5 Incubation

From a purely economic standpoint, today's egg or poultry meat producer has little reason to hatch his own stock. He can rely on specialist hatcheries to supply him with day-old chicks more cheaply than he could produce them himself, taking into account the expertise and equipment required. The same thing does not apply to the small-scale breeder, however, or the owner of a limited number of ducks, geese or quail. Even the producer with a small laying flock, if he is using pure lines or crosses, or living in a remote area, may find it necessary to hatch his own replacements.

Incubation can be viewed on two planes therefore; that of the self-sufficient poultryman, using broodies or incubators, or a combination of both, to generate his new stock, and that of the business-orientated hatchery geared to the incubation of hundreds of thousands of broiler, layer or turkey eggs to meet the needs of big commercial customers.

The systems may differ, but the same fundamental principles apply. After all, an embryo will only develop if given the right conditions of warmth, humidity, movement and ventilation, regardless of the size of the enterprise. There is even a certain amount of overlap in the way the systems are run. A breeding flock farmer does not have to be in a very big way of business himself, for example, to be a professional supplier of hatching eggs to a large hatchery. Plenty of hybrid parent flocks consist of little more than a thousand birds, managed by a farmer who has an agreement to supply eggs under terms laid down by a breeding company.

Egg selection

Whatever the eventual method of incubation, choice of the hatching egg is the important first step. Successful hatchability depends on a chain of circumstances, starting with the breeding flock. If they are healthy birds of optimum age with a good fertility record (as discussed in Chapter 4), and receiving a properly balanced breeder's ration, the proportion of fertile eggs should be high. For the next phase of selection the incubationist relies on what he can see.

Are the eggs of the right size?

Not too little and not too much is the maxim. Dealing with species from the quail to the goose, it is impossible to be dogmatic about precise weights, but even with the fowl recommendations range from 53 g to around 64 g ($1\frac{7}{8}$–$2\frac{1}{4}$ oz). The general rule is to choose eggs which are of the commonest size laid by the breed in question.

Are they the correct shape?

Misshapen eggs may hatch sound chicks, but there is the chance that they will not, or that the characteristic for badly shaped eggs will be inherited. Wherever possible, reject them.

Are they sound?

'Candling' is not always carried out with new-laid eggs, but it can be useful in picking up otherwise unseen faults. Originally done, literally, with a candle, it simply involves rotating the egg gently in front of a concentrated light, preferably in darkened surroundings. Torch-shaped candlers can be easily bought for small numbers of eggs, and more elaborate versions, which will illuminate whole trays of eggs at a time, are used by large hatcheries. Today's commerical practice is to make just one candling, when eggs are transferred from hatcher to settler, however.

Such candling shows up cracks or weak spots in the shell, blood or meat spots or double yolks, all of which naturally downgrade an egg for hatching. At the same time it gives an indication of possible breeding flock problems, by showing up infertiles after only a few days of incubation. A large number of 'clears' which show no germinal development when broken out and examined, point to infertility of the parents. By the time an egg is laid, having been fertilized at the upper end of the reproductive tract, the blastoderm has already become a disc of several hundred cells.

Are they clean?

Shells are porous and nest areas team with bacteria—especially in the case of water-birds—so it makes sense to exclude any egg which is more than usually soiled. Penetration of the shell pores by micro-organisms can lead to moulds, 'rots' and possibly 'exploders', which are eggs which burst and spread infection to all their surroundings. If dirty eggs are particularly valued for any reason they can be cleaned, as described in the passage on hygiene in this chapter, but it should be done without delay.

Collection and storage

Hatching eggs should be removed from the laying area as soon as practicable. Breeding flock farmers will often collect from the nests three to five times a day.

The eggs are generally put on to fibre trays known as Keyes trays, carefully stacked on low, wheeled trolleys and taken to a room at the end of the poultry house where they are transferred, still on their trays, to cardboard boxes, ready for despatch to the hatchery. But this involves a fair amount of handling, both at the farm and the hatchery, and increases the risk of cracking, so the main incubator manufacturers have developed plastic egg trays and special trolleys which fit their big, walk-in machines. Now eggs can be placed straight in the plastic trays and loaded into the incubator trolleys, which are brought empty to the farm. When the transport arrives from the hatchery, the trolleys can be wheeled aboard and taken to the hatchery with a minimum of fuss.

Incubation does not progress automatically as soon as eggs are laid. They can stay in a state of suspended animation without harm for several days, providing they are cool. The triggering factor is warmth. Embryonic development starts at something above 20° C (68° F), so, for safety, holding temperatures must clearly come below this. On the other hand they must not be so low as to create a risk of chilling. Most incubationists aim for a consistent temperature in the range 10–16° C (50–61° F), the optimum being 13° C (55° F).

At hatcheries the holding room is quite often windowless, well insulated and equipped with proper cooling apparatus. Domestically, even in hot weather, it is usually possible to provide the right conditions for small numbers of eggs over short periods, remembering that they need not only a consistent low temperature but also adequate ventilation and a relative humidity of 70-80% to stop moisture evaporating too quickly.

Relative humidity

It is useful to understand the meaning of relative humidity, often abbreviated to 'RH', since apart from its importance throughout incubation, it also has a bearing on troublesome environmental conditions, such as wet litter, later in the life of stock.

Measurement of relative humidity is made by comparing the readings of a conventional thermometer, the 'dry bulb', and a 'wet bulb' thermometer, in which the bulb is surrounded by moistened material.

Evaporation creates a lower temperature at the wet bulb thermometer and the only time the two thermometers will give the same reading is when the atmosphere is completely saturated and evaporation stops. This is 100% RH.

As the air's temperature rises, however, so does its ability to hold moisture. If the temperature is high but the atmosphere is naturally dry, the dry bulb will record the high temperature but the wet bulb will have a high rate of evaporation and show a correspondingly low temperature.

More moisture in the air, or a lower temperature with the same moisture level will narrow the difference between the two thermometers. In other words, the closer the relationship of the thermometer readings, the higher the RH.

These ratios can be charted, and generally a hygrometer—a wet and dry bulb thermometer—will be supplied with a table showing the percentage of relative humidity at given wet and dry bulb readings.

At 13.3° C (56° F), close to the optimum storage temperature, for example, a wet bulb reading of 10.6° C (51.2° F) will give 70% RH. If the wet bulb registers 11.5° C (53° F) it indicates an RH of 80%—still acceptable, but moving towards over-humidity.

Too much moisture has to be guarded against, especially in warm conditions. At the 'dew-point', condensation on the shell simply supplies a free swimming-pool for potentially dangerous micro-organisms.

In practice never ignore the value of the hygrometer. Keep the instrument clean and check that the wick is kept moist, preferably with distilled water, since this contains no mineral deposits which might distort the reading.

To control levels of humidity the well organized hatchery or farm holding-room will have humidifiers installed, but these are hardly a worthwhile expense for the small operator. Open pans or trays containing water, placed in the storage area, will be sufficient where humidity is low.

Apart from a pre-incubation period of twenty-four hours when the eggs can be gently acclimatized at around 20° C (68° F), storage time should be as short as possible; ideally, less than a week.

In a 1976 experiment, Christine M. Mather and K. F. Laughlin of the Poultry Research Centre, Edinburgh found that the hatchability of eggs stored for fourteen days was 19.5% lower than for non-stored eggs and the incubation period was 13.4 hours longer, equivalent to one hour per day of storage. These results, which accord well with earlier work by other researchers, are fair warning against extended storage. So although experimental work with pre-heating eggs, storage in plastic bags or in inert gases shows some promise, it is best to play safe and keep down holding times.

Although turning of eggs is essential during actual incubation, to prevent the embryo from floating upwards and adhering to the shell membrane, it is not vital during limited periods of storage. Nevertheless, for any period over a week it becomes important, and nothing is lost by practising it from the outset.

If eggs are stored flat they can be rolled through 180° daily, reversing the direction of roll each day. If they are in Keyes trays, the trays can be tilted to about 45° in each direction. Note, incidentally, that eggs for incubators are trayed with the small end down, to give the embryo the best chance to reach the air space just preparatory to hatching, and to expose the maximum area of shell for ventilation. It is a good idea, therefore, to cultivate this habit of 'pointed end down'.

Natural incubation

For the small-scale poultry enthusiast there is a pleasant sense of old-world satisfaction to be had from a mother hen and a brood of chicks. Nor is it entirely sentimental. The requirements are quite simple: broody hens, suitable accommodation, food and water. In return the results can be good. Even now some moderate-sized operations employ flocks of broodies, the main reason they are becoming rare being the tremendous growth of the industry, which demands many thousands of chicks at one time. It has to be recognized also that there is a substantial risk of cross-infection between mature birds and chicks.

Because of the way the industry has moved, acquiring suitable broodies might be the small poultryman's first problem. Egg producers do not want 'non-earners' about the place so broodiness, being an inheritable characteristic, has been largely bred out of laying birds by selection over a number of years. As a result commercial layers make poor broodies and it is necessary to look elsewhere for suitable sitters. This is where heavier exhibition breeds and crosses which have not lost the maternal instinct come into their own. Silkies, for example, which look particularly uncommercial with their delicate, floating plumage, have an excellent reputation as mothers, and good sitters will often be found to have Silky blood in them.

Early signs of broodiness are not difficult to recognize. The hen shows reluctance to leave the nest, makes distinctive clucking noises and tends to ruffle her feathers if approached. Watch her over two or three days and provide her with food and water as though she were sitting properly. If the broody behaviour persists, she will probably make a good sitter.

When there are no outward signs of broodiness, the condition may be induced. Hens, like other birds, lay eggs in clutches and if the eggs are not taken away day by day, they will pause when a manageable number have been laid, to incubate them. It is sometimes possible, by leaving upwards of half a dozen eggs in the nests of prospective sitters, either real or dummies, a fortnight or so before you need broodies, to bring out the natural instinct.

Accommodation need not be elaborate. It should provide a quiet haven, dry and well ventilated, but free of draughts or unwelcome visitors, including other, non-sitting, chickens. For a small number of sitters a garden shed or similar outbuilding, preferably with an earth floor, can provide the right surroundings, or a timber coop can be made up. Basically all that is needed is a box with interior dimensions of about 38 cm (15 in) square and approximately the same height to hold the nest, and an adjoining area, which can be separated from it by a sliding shutter, of say 61 cm × 38 cm (24 in × 15 in) to give the hen access to food and a dust bath. The whole thing stands directly on the ground. The nest box should have a sloping roof which can be detached and the 'run' area can be either slatted or wired in at the the top. To stop rats or mice, the floor should be of 12 mm ($\frac{1}{2}$ in) mesh wire-netting.

Silky hens make good
mothers

An alternative is a simple ark in the form of an equilateral triangle,
both sides and the open base being 61 cm (24 in) across, and the whole
structure 122 cm (48 in) long. In this case it is probably easiest to make
the end wall detachable, and again a section of about 61 cm (24 in) can
be left open at one end to provide a run, though these open sections
should be given temporary covering in wet conditions.

Wood, exterior hardboard, metal, glass fibre and a host of other struc-
tural materials can be used, providing due attention is paid to ventilation
and weather-proofing. This is particularly important with impervious
substances, which can build up condensation. Holes of about 19 mm
($\frac{3}{4}$ in) diameter should be drilled at points where they will provide a
reasonable air-flow without allowing water in.

In a shed, the nest base is best made from turf with some of the earth
scraped away from its underside so that the centre falls in slightly, laid
over earth in a shallow box. The same arrangement can be used for outside coops, though here the turf, cut to the size of the nest box, may
be set directly on the earth floor. In both cases a bed of fresh, broken
straw is then set on the turf.

Dry, fine ashes, mixed with good insecticide, such as pyrethrum
powder, are put down just outside the nest area, to make a dust bath,
and all is then ready for the broody.

Mites and other vermin can make a hen a very restless sitter on her
long vigil, so it is as well to give her a dusting also with insecticide before
introducing her to her new quarters, though she should be handled with
a minimum of fuss.

The transfer can be made at any time of day, but is probably least
disturbing at nightfall, which gives the hen the night and the whole of

the next day to acclimatize, with a few 'sample' eggs to keep her company. By the following evening she can be allowed off the nest—it may be necessary to lift her—given food and water and the opportunity to relieve herself in the pen area, and the proper clutch placed in the nest. She can then go back to the nest to discover the eggs for herself.

Do not overload a broody. A fowl of average size will cover fifteen hen's eggs or twelve turkey eggs. Depending on their breed, she could also sit twelve duck eggs or six goose eggs. Bantams will take about six of their own eggs. But in every case it is better to err on the side of too few than too many.

Once the hen is established on the nest little further need be done, apart from supplying regular food and water. Indeed, it seems almost an impertinence to suggest that one can teach a bird to sit eggs, but it is true that their consistency varies, and some attention during the nesting period will help to assure maximum hatches. Even daily turning of the eggs is sometimes recommended, but they are shuffled sufficiently by the sitter for this not to be necessary.

The best feed is grain, such as a mixture of maize or corn and wheat, with no soft mash or greenstuff, which can lead to soft droppings and fouling of the nest. Only one feed a day is necessary and if possible this should be supplied to a set routine, either in the evening or the morning. Adequate fresh water is important. At feeding time the hen can be lifted or encouraged off the nest and allowed a stretch in the feeding area.

The eggs can safely be uncovered for up to half an hour and in this time the nest can be checked for any dirty straw or broken eggs. In either event, if the eggs are soiled they should be carefully cleaned with a stiff brush or wire wool, or washed in water at 30° C (100.4° F) and new nesting material put in the nest. Do this just before the hen returns to the nest, to avoid chilling due to evaporation.

Quite often, when allowed access to the nest, the hen will return of her own accord, but if not, especially in the first few days of incubation, she must be put back in the confinement of the nesting box. Later she will probably adapt to the routine and can be allowed to come and go freely between the nest and the feed.

Occasionally a sitter becomes restless—infestation will cause this—and stands up on the eggs. If this is noticed before they have become cool, they can be rescued and kept in a hay-box until a replacement broody can be found. Even after forty-eight hours kept in this way, eggs have been known to hatch.

Following about a week of incubation, it can be helpful to the small operator to candle the eggs. By this stage, those developing normally show their dark contents and a network of blood-vessels. The contents are mobile when the egg is turned. Dead germs (where progress ceased at an early stage of embryonic development) now frequently show up with a characteristic blood ring, lack of blood-vessels or a fixed position where they have stuck to the shell.

Where several broodies are being used, it pays to sort out the duds now, since removing them and redistributing the others into new clutches makes the best use of available hens. For example, by setting three broodies simultaneously, two may be able to cover all the fertiles detected after seven days and the third re-set with a fresh clutch. When the chicks hatch, the same principle can be applied, two hens taking on all the chicks while the third can be re-set on another clutch. The infertile eggs can be hard-boiled and used in the chicks' mash during the first few days after hatching.

Occasionally a hen may need help in dry conditions to supply sufficient humidity for the eggs. If the nest earth shows signs of drying out, particularly in the last week of incubation, about 500 ml (1 pt) of warm water should be poured on the ground surrounding the sitting-box each day, and it may be necessary to spray the eggs lightly with warm water just before the hen resumes sitting.

Artificial incubation

While the broody hen involves some labour in looking after her daily needs, she does have the happy knack of providing exactly the right temperature and general environment for hatching eggs, plus a certain amount of post-natal care for the chicks when they hatch. By contrast, artificial incubators have turning mechanisms to seize up, fuses to blow, thermostats to malfunction, heaters to catch fire, humidifiers to dry out, and a total indifference to emerging offspring. Yet small machines are an important aid, even to operators who prefer broodies, while without the large incubators there would be no modern poultry industry.

The point is that, while the incubationist should regard every machine with a healthy suspicion—more so if it is secondhand—a high proportion of failures are due to 'pilot error'. No equipment should be blamed for lack of routine maintenance, neglect during operation or the owner's forgetting to read the manufacturer's instructions. Check, check, check conditions throughout the incubation period.

Incubators fall into two main categories; still-air or flat machines and forced draught or cabinet machines. The former are the type best suited to the needs of the back-yarder, with capacities ranging from as few as five hen's eggs (really only suitable for school demonstration) to a few hundred. Cabinet incubators, including the so-called mammoth machines, range to infinity these days, with modules than can be added as a business grows.

The principle of the still-air incubator is that the eggs—generally lying flat on a tray—are warmed by radiant or convected heat. A common type employs an adjacent heating chamber holding either hot air or water warmed from a heat source outside. The heat may be supplied by oil, bottled or town gas, or electricity. Little difference is to be found in the results from these machines, although the tank incubator containing water is sometimes considered better for duck eggs, and retains its heat longer if anything cuts off the heater. On the other hand the hot-air chamber needs no filling and reacts more quickly to temperature control.

Choice among small incubators runs from this type, generally totally enclosed with just a small observation port, to 'see-them-hatch' machines with a transparent lid, much favoured by biology teachers. If a purchase decision must be taken between makes, probably the best rule is to work out the cost per egg of the machine you have in mind.

Before setting eggs, however, the same procedure applies to every machine. Make sure it is thoroughly clean, disinfected and working properly. Have it running for at least two or three days without eggs to ensure that correct temperatures are maintained. Check the thermometer by reading it against a second one, placing them both first in cold water, then at about 39.4° C (103° F) and finally in hotter water, but take care that it is not at a temperature that will burst the thermometer.

Care should be taken over siting an incubator. It should stand on a level and, if possible, solid floor, away from any window which might allow direct sunlight to play on it. Good ventilation without draughts is necessary.

If all has gone well in the test period, the eggs can now be set. Here the manufacturer's instructions should be followed closely with regard to capacity. It is sometimes tempting to put in a few extra eggs, on the grounds that some infertiles will be removed when the eggs are candled on the seventh day. Resist that temptation, and also any inclination to put more eggs in after the hatch has started.

Some machines are equipped with mechanical turning devices, in others the eggs must be turned by hand, but in either case it helps if the shells have been marked in some way. With pedigree birds, a coding

Maran chicks hatch in a small incubator. Letters on shells can act as identification and a guide to egg turning

of the parentage may already have been written on, or the date of collection. Failing this type of information, a simple mark such as an 'X' is a sufficient guide to turning. Where eggs are laid flat they should be turned through 180°, in alternate directions, at least every six hours and more frequently if possible. The critical period is in the early stages of incubation—for fowls from day three to day nine—though turning should continue up to a day or so before hatching (see chart).

It should be a matter of daily routine to inspect, and if necessary replenish, fuel tanks on oil-heated machines, to pay attention to humidity levels and to see as far as possible that all eggs received the right temperature.

Temperature requirements for different types of poultry are very similar, so that turkey and duck eggs, for example, can generally be incubated in machines designed with chickens in mind. Goose eggs, because of their relatively larger size (around 200 g (7 oz) compared with the 58 g (2 oz) average for fowls) may not be so easily accommodated in some machines. It is, however, highly desirable to have eggs of only one species in a still-air machine at any one time.

Incubation requirements for poultry species

	Fowls Standard	Bantam	Turkeys	Ducks	Muscovy ducks	Geese small	large	Japanese quail	Guinea fowl
INCUBATION PERIOD (DAYS)	21	20	28	28	35	30	34	17	28
TEMPERATURE* STILL AIR MACHINES	39.4°C (103°F)	39.4°C (103°F)	38.1–38.6 –39.2– 39.4°C (100.6– 101.5– 102.6– 103°F) in 1st, 2nd, 3rd 4th weeks	38.6°C (101.5°F)	38.6°C (101.5°F)	39.2°C (102.6°F)	39.2°C (102.6°F)	38.75°C (101.75°F)	38.75°C (101.75°F)
FORCED– DRAUGHT MACHINES	37.6°C (99.75°F)	37.6°C (99.75°F)	37.5°C (99.5°F)	37.5°C (99.5°F)	37.5°C (99.5°F)	37.2°C (99°F)	37.2°C (99°F)	37.6°C (99.75°F)	37.6°C (99.75°F)
RELATIVE HUMIDITY	day 1–19 55–60% hatcher 70%	day 1–19 55–60% hatcher 70%	day 1–24 65% day 24–28 75%	day 1–24 70% day 24– pipping 60–65% hatcher 70%	day 1–32 60% hatcher 75%	day 1–27 70% day 27–30 75%	day 1–30 70% day 30–34 75%	day 1–14 55–60% day 14–17 70%	day 1–25 55% day 25–28 70%
TURNING (UP TO AND INCLUDING)	day 18	day 18	day 24	day 24	day 31	day 26	day 30	day 15	day 24

Note: These figures are a guide to optimum conditions, but as experienced incubationists know, variations around the mean can be quite wide, depending on circumstances. Incubation periods, for instance, may be longer or shorter by days, so operators must be prepared to make adjustments based on personal observation.

* Follow incubator manufacturer's recommendations where they differ from these figures.

While the ideal temperature at the centre of an egg is something close to 37.8° C (100° F), in practice its external environment varies, of course, and even under a hen the upper shell surface will be about 39.4° C (103° F) while the lower side will be cooler. The still-air incubator tends to duplicate this situation, the thermometer bulb being placed near the upper surface of the eggs, where it registers a higher temperature than that below the eggs. This type of machine is therefore run at 39.4° C (103° F) for fowl's eggs, with just marginal variations for other species, as can be seen in the chart. With big forced-draught incubators, however, uniformly warm air actively circulates around the trayed eggs, below as well as above, and temperatures are kept closer to the optimum, at 37.8° C (100° F).

Unlike large cabinet incubators, humidity is not easy to control precisely in flat machines, though humidifying devices are often fitted, and these should be topped up with water regularly, according to the maker's instructions. Sufficient moisture is particularly crucial in the hatching phase. Three to four days before eggs are due to hatch they should be damped with a warm wet cloth when they are turned. Duck and goose eggs benefit from a daily sprinkling with water at incubation temperature during the first two weeks, then twice a day. Just before hatching, a small sponge soaked in warm water can be placed in each corner of the egg tray.

Humidity control is apt to be a little haphazard in small incubators, which do not generally have a hygrometer, but even in large machines the hygrometer must not be regarded as an infallible guide to conditions inside the egg. It tells you the humidity levels in the incubator, and these have a vital bearing on the speed at which eggs evaporate their moisture during incubation, but there are some variables to take into account. Eggs of one batch may be of a different size or have more porous shells than another, which will affect their rate of moisture loss. So the ideal RH for the first batch may be a little too high or too low for the next.

Evaporation is a natural process in incubation, eggs shedding about 12% of their moisture during setting. How efficiently they do this has a profound effect on the timing and success of the hatch. If humidity levels have been too high, slowing down evaporation, it will tend towards premature hatching of sticky chicks. Too little moisture will result in failure of a high proportion of the chicks to break out, with those that do being delayed.

It is not enough simply to apply recommended RH levels therefore. Moisture loss should be monitored as well. This can be done by sampling eggs at regular intervals and inspecting the air space with a candler. Charts can be obtained showing the optimum space at different stages of incubation, but a number of eggs should be checked, to allow for individual variations. Alternatively samples can be weighed, and this is the policy adopted in many large hatcheries. A tray of eggs will be weighed initially and then at four-day intervals, to see that they are on course for the optimum 12% reduction by the eighteenth day, if they are chickens' eggs.

Hatching records fall into two categories; those which keep the incubationist informed on details of the hatch, and those which give a running account of procedure day by day. Both can be easily set out on a chart. The former will note the source of the eggs, the date they were set and the number. It will show the number rejected at first candling, the number of 'clears', how many were removed at the second candling and the number of dead-in-shell. It will also indicate the number of chicks hatched, as a percentage of eggs set, and there may be space for additional remarks. The running record, on the other hand, will show incubator temperature, humidity and/or egg weight loss, and frequency of turning. It may give a check on temperature and conditions in the incubator room and, in large hatcheries with the facilities, possibly detail any changes in carbon dioxide concentration.

Such records can be invaluable if a hatch fails unpredictably. It may mean checking back over three weeks or more of work to find the fault, but a chart will simplify the job and help to ensure that the next hatch is not spoiled in the same way.

Hatchery practice

Commercial hatcheries are a world apart from the domestic hen, of course, yet they perform the same basic function. The principal difference is one of scale. Consider this specification for a hatchery built in 1977:

Ground area: 1040.5 sq m (11,200 sq ft)
Egg capacity: 308,000
Facilities: Egg handling and storage areas
           Incubation, hatching and chick handling rooms
           Delivery bays for four vehicles

The metal-framed structure is heavily insulated and lined on the inside with asbestos board sealed with epoxy resins. Porcelain-treated ceiling panels provide a non-porous interior for easy sanitation. Gas-fired heat transfer units provide each room with its own environmentally conditioned air supply and humidification is automatic.

This particular hatchery is owned by Shaver Poultry Breeding Farms Ltd, and operates in Canada. It is not exceptionally large and is just one of a number run by this international breeding company, but it illustrates the broad pattern of production today, with such hatcheries, quite often owned by breeding companies, producing practically all the world's broilers and the majority of the layers and turkeys in the developed countries.

Where a hatchery is not owned by a breeding company, it may hold a franchise with one or more breeders, permitting it to hatch their stocks. Under either arrangement hatcheries tend to specialize, keeping the variety of breeds they produce to a minimum. In this way they can standardize their management approach for the very large numbers of birds they will hatch in a year.

As far as possible, newly-built hatcheries are situated away from farming and other activities to reduce contamination problems, and hatching eggs are brought in from supply farms often scattered over quite a wide surrounding area. Depending on company policy, these farms may be wholly-owned sites managed by company employees, or privately owned and operated under a contract or written agreement with the hatchery, as mentioned at the beginning of this chapter.

Under a typical arrangement the farmer will supply housing, fuel and management facilities in exchange for the breeding birds, veterinary attention, feed and payment for the hatching eggs produced. Occasionally financial credit may be offered. Every week or so a company fieldsman will call to check on progress, weighing the birds during the rearing period, for example, and offering advice if required.

Eggs collected for the hatchery are either code-marked or delivered with a consignment note identifying the farm of origin. In this way, with the records kept at the hatchery, faults arising on any site through, say, a nutritional deficiency, can be quickly traced back and remedied.

A hatchery office is a busy place, for in addition to maintaining the routine statistics, looking after wages and so on, it must chart a programme to keep its incubators stocked to economic capacities, supervise transport for egg collection and chick deliveries, and formulate a long-term sales plan.

Large incubators

Contrary to popular belief, the artificial incubation of large numbers of eggs together is really nothing new. Ancient Egyptians were using 'walk-in' type incubators, built in mud brick and heated by burning camel dung, before Britons were out of woad. Moreover, some of these establishments could hold up to 90,000 eggs and return hatches of over 70% from the fertiles set. Incredibly, incubators in much the same design were still operating up to the late 1950s, and a few may still exist. China and other areas in the Far East also have a history of very early developments in artificial incubation.

A big incubator, when loaded, is like a living organism itself, made up of thousands of cells—the eggs. These take in oxygen, produce carbon dioxide and, in the later stages, generate their own heat. If anything upsets the delicate balance for long, they die. Therefore everything is geared to maintain that balance.

Forced draught machines employ paddles or fans to circulate the warmed air among the eggs, ensuring that no cold spots develop and that correct ventilation levels are maintained. These devices must operate at the right speeds, according to manufacturers' instructions, and they need to be checked perhaps two or three times a year to see that they have the right rpm. If ventilation is low, heat and carbon dioxide will build up and moisture loss from the eggs will accelerate.

Experts recommend a carbon dioxide level of 0.4% or slightly less for best incubation results. Fresh air contains about a tenth of this. Special gas 'sniffing' devices can be used to detect a build-up. But if ventilation stops in an incubator containing thousands of eggs, the gas accumulation is rapid and at concentrations over 2% embryonic deaths will be widespread, if not total. Auxiliary generators are important in a hatchery.

Since hatching requirements are different from those during the main 'setting' period of incubation it has become widespread practice in major hatcheries to have separate setters and hatchers. This opens the way for some economies. After thirteen to fourteen days the chick embryo is generating its own heat which, with the large numbers of eggs involved, must be dissipated. By filling a third of the setter each week, embryos at three stages of development are involved, some requiring heat while the others are producing it. In this way the cost of supplying artificial heat can be kept to a minimum.

To pack eggs into the smallest reasonable space it has long been the policy to put them on end in trays for forced draught machines. Older incubators still take trays of wood and wire in which the eggs are loaded 'shoulder to shoulder', but there is now a move towards plastic trays, generally holding 132 eggs, each with its own compartment, which gives better protection against breakages and fits into modern handling techniques. Whatever the tray system, however, it means that eggs cannot be rolled, as in flat machines, but must be tipped in alternate directions.

Some ingenious devices have been developed for egg turning. Trays

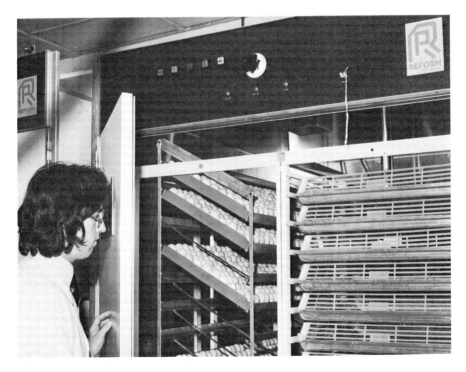

Eggs are turned regularly and automatically by tipping the trays in a large incubator

may be placed directly into racks in smaller cabinet incubators, but they are loaded into trolleys and wheeled into the big, walk-in versions. There the trolleys are linked into the incubator's turning mechanism and the trays tilt backwards and forwards in the trolleys.

In all large incubators it is an automatic process which usually takes place every hour. Most turning devices tilt the trays through 90°, which is 45° in each direction, but this is debatable. I have heard reports of better hatching results for turkeys attributed to a machine which tips them 60° in each direction. The important thing, once again, is to make sure the mechanism is functioning properly at all times.

Hygiene

There can be no finer site for multiplying micro-organisms than a fertile egg. Pharmaceutical companies recognize this. They use them in their thousands for vaccine cultures. But the conditions which these companies have turned to man's advantage can also work the other way, of course, and promote unwanted organisms in a hatchery which reduce the hatch or, worse, spread infection further afield. Since millions of day-olds are now distributed, nationally and internationally, every week, this puts a heavy onus on hatcheries to keep up effective sanitation programmes. In general they do, the pattern only varying slightly in detail from place to place.

As far as possible, the interiors of hatcheries are isolated from the outside world. Eggs, people, equipment and even air can only enter by recognized routes, and a one-way system ensures that nothing 'dirty' is allowed to pass back to a 'clean' area. Shells and chicks from one hatch, for example, will have no contact with eggs newly arrived at the hatchery.

Starting at the supply farm, every egg is checked when first collected and any obviously soiled are separated from the clean ones. Depending on hatchery requirements, dirty eggs will either be cleaned before setting, or used for some purpose other than hatching. Where water is used for washing it must be at a slightly higher temperature than the egg, to prevent its being drawn in through the pores by contraction of the egg contents. It should contain a germicide at the dilution recommended by the maker.

Large hatcheries may be equipped with mechanical washers through which the eggs pass as a matter of routine, but not every hatchery management agrees on the value of washing. It is, after all, an extra process, and not essential if the breeding farms are doing their job in producing nest-clean eggs. Also it must be done properly, with correctly designed machines and no re-cycling of the cleaning water, if cross-contamination is to be avoided. Against this, supporters of the washing argument often point to improved hatches and less dusty conditions in the hatchery. One turkey hatchery which washes eggs and dips them in a germicidal solution has recorded improvements of 2% in hatchability, and mortality in the first sixteen weeks cut from 4% to 2.5%.

Whatever traying system is used for the eggs, the same trays will be returned to the same farm as far as possible. Fibre trays will be replaced if they are not clean. Plastic trays will be washed and disinfected between each load, usually at the hatchery.

If the farm is equipped with its own fumigator the eggs may be fumigated on the spot. If not, they will be treated on arrival at the hatchery. Formaldehyde gas, although unpleasant and potentially hazardous if carelessly handled, is an efficient and economical disinfectant, and the most widely used.

Modern hatchery design is such that incoming eggs can only go from the holding room to the incubators by way of a fumigation tunnel or chamber. They are transferred into setter trays and loaded into trolleys in the storage area—unless, using the latest methods, this has been done at the farm. The trolleys are then wheeled into the fumigation chamber, the gas introduced and the doors sealed.

Formaldehyde gas can be produced in several ways. The heat system is popular in farms and hatcheries because it employs paraformaldehyde, a chemical easy to store and handle which gives off the gas when heated on a simple electric plate. It is supplied in the form of 'prills', used at a concentration of approximately 170 g per 28.3 cu m (6 oz per 1,000 cu ft) in the fumigation chamber. Since the gas is dry, humidity levels are normally raised to 70-80% by using a spray or aerosol generator just beforehand, and it works best at 21°C (70°F). Heating devices capable of holding several pounds of prills can be obtained, and if they are fitted with time switches, the process can be automatic.

Some people prefer to use formalin—a solution of formaldehyde in water—which they add to potassium permanganate crystals. Note the order: formalin is always added to permanganate, never the other way round. Reaction between these two chemicals is violent, bubbling and releasing clouds of gas vapour. In this case 80 ml (3 fl oz) of formalin to 57 g (2 oz) of potassium permanganate, for every 2.83 cu m (100 cu ft) of space in the chamber, is a usual concentration.

Another effective method of using formalin is as a mixture, in equal parts with water, dispersed by aerosol. About 140 ml (5 fl oz) of formalin and 140 ml (5 fl oz) of water is sufficient for 28.3 cu m (1,000 cu ft).

As a surface disinfectant, formaldehyde is harmless to fragile materials yet a killer to most bacteria, even in the presence of organic matter. For this reason it is widely employed in different parts of the hatchery, but the same precautions must be taken in all circumstances. Wear a respirator and eye shield because exposure to the concentrated gas can set up strong irritation of the nose and eyes. If there is any danger of getting splashes of formalin on the skin, rubber gloves and protective clothing will be necessary, and wherever fumigation is taking place, doors should be locked or warning signs hung up.

Although eggs easily withstand their thirty minutes exposure in the initial fumigation chamber, they must never be fumigated between the twenty-fourth and ninety-sixth hour after setting. By the same token, although evaporation of formalin solution from a tray or cup in the hatcher is not unusual, chicks should never encounter the gas in concentrated form.

Setters will naturally be cleaned down thoroughly and frequently, a useful tool here being the high pressure spray and sanitizing fluid. This will generally be followed up by fumigation. The incubator room is also regularly scrubbed down, sprayed and disinfected.

Periodically, in most hatcheries, veterinary or health inspection teams take a 'bug-count' in strategic parts of the building, placing slices of agar gel on the floor, walls and working surfaces and examining the subsequent cultures they build up for any pathogenic organisms. Hatchery staff may also be checked.

Nobody should be able to enter a working hatchery without first going through the hygiene procedure—the risk of carrying contaminants on hands or feet is too great. All employees will therefore have their entry room, with adequate wash basins, germicidal soap and paper towels. Here they can change into working overalls and footwear to be used only in the hatchery. At least one company supplies different coloured boots to be worn in different sections of the building, as a check that employees do not stray inadvertently from, say, the hatching room to the egg handling room.

Research is constantly going on to find cheaper, more potent, less toxic, faster acting or more easily stored disinfectants, and the product line is a long one, but basically only a few active ingredients are involved, each best suited to a certain job. Quaternary ammonium compounds, abbreviated to 'quats', are probably most widely used for general hatchery hygiene. Broad spectrum disinfectants—those which kill a large cross-section of infective organisms—are made from these. They act fast and are quite effective deodorants. They can be used in fogging equipment as well as direct application.

It is important when using disinfectants to ensure that they are compatible with detergents or other cleaning compounds, otherwise their effect can be neutralized. Quite often manufacturers formulate them with a suitable cleaning agent.

Others in the disinfectant range include the iodophors, which combine elemental iodine, itself a potent microbicide, with a surface active agent or 'surfactant', to make them easily soluble in water; the synthetic phenols and the chlorine-based compounds.

In the microbe war, airborne invasion is always a threat, of course, and to combat this up-to-date hatcheries use air pressurization, making it slightly higher in the clean than the dirty areas so that air-flow cannot carry harmful contamination into disinfected regions. Modern incubators commonly have their own filtered ventilation direct from outside the building.

Undoubtedly the infection peak in any hatchery comes at hatch time, when live chicks, shell debris, dust and fluff are around in abundance. Normally eggs are transferred from setter trays to hatch trays at pipping stage (when the birds have broken the first small hole in the shell). If the move can be made well before pipping, so much the better. Candling allows the obvious non-hatchers to be removed, and the emptied setter trays can be taken away to be soaked in disinfectant and pressure hosed to remove any organic matter before being returned for the next batch.

Often a pebble-surfaced or coarse weave cloth is placed in the hatching trays before the eggs are put into them. A recent development is a plastic hatching tray with a textured floor. Such surfaces enable the chicks to gain a good footing and perhaps help to avoid some spraddle-legged cripples.

As soon as hatching has been completed and the chicks have been packed into clean boxes for despatch, the hatcher trolleys and trays can be removed to the wash room for disinfection and hosing, while hatch debris is disposed of, preferably in closed containers. The cleaning team will also move in to decontaminate the hatcher, vacuum cleaning to remove down, power spraying and disinfecting, before giving the machine a final fumigation with all inlets and outlets closed.

Trouble shooting

A good hatch will be concentrated into a relatively short period of about twenty-four hours and will result in sturdy, bright-eyed chicks with a good covering of fluffy down and plenty of life, emerging from virtually all the eggs in the hatcher. Anything less spells trouble, not only for the current hatch, but for succeeding ones, if the fault is not rectified. Clues can usually be found in a stage by stage diagnosis.

Infertility or embryonic mortality?

Look to the first candling at five to ten days of incubation. Rejects at this stage will be mainly clears or blood rings. Samples should be broken out to check. A high proportion of clears suggests a fertility problem calling for investigation among the breeders. Suspect an aging flock, the wrong ratio of males to females, preferential mating in pens, disease, parasites or incorrect nutrition. Tips on feeding of breeding birds will be found in the chapter on nutrition.

Blood rings indicate early embryonic death, not infertility. They can be caused by dietary deficiency in the parent stock but are more likely to be due to mishandling of the egg in the early stages. Were the eggs stored at the recommended temperature and humidity and for the shortest possible time? Were they turned? Was the correct amount of fumigant used? Are you sure setter temperatures were accurate? Apart from misleading thermometers or faulty thermostats, a fluctuating power supply could be the culprit.

Large numbers of dead-in-shell?

Seen at the candling stage, when eggs are transferred to the hatcher, this implies several possibilities during the setting period. The field can be narrowed by inspecting the embryos. If a lot of deaths appear to have occurred around the mid-term period (between days ten and fourteen in fowl embryos) faulty nutrition may be the cause. Tiny nodules covering the skin surface—clubbed down—indicate a riboflavin deficiency in the parents' diet, for instance.

A bird can deviate from the normal pattern of embryonic development and still hatch successfully. Generally, at hatching, the bird is lengthwise in the shell, with the beak beneath the right wing and its tip towards the air cell at the blunt end of the egg. At the World's Poultry Congress in 1948, however, Marshall classified several malpositions. Those recognized as affecting hatchability include 'head in small end of egg' (often as a result of eggs being set blunt end down); 'head to the left'; 'body rotated'; 'feet over head' and 'embryo across egg'. Shaking or other rough handling of eggs, some lethal genes or insufficient turning can all contribute to malpositions.

Again, incorrect temperature and/or an accumulation of CO_2 will lead to dead-in-shell. Egg infections are often self-evident. Hygiene at the hatchery or the supply farm will need to be tightened up.

Mortality at pipping?

Humidity levels are the first check-point here. RH should be 70% or slightly over for poultry species when pipping commences, though it may be dropped temporarily to dry the shell and improve the gas exchange as the embryo prepares to change over to lung breathing. Check that hatcher air intake and exhaust vents are adjusted accurately, fan speeds are correct and humidifiers functioning properly. The hatcher should not be opened frequently.

Slimy-looking offspring—as opposed to the natural wetness of all newly hatched birds—call for a check on humidity during setting. Was it too high? The record of wet bulb readings should tell you, but ensure that the eggs were turned properly, particularly in the first two weeks.

If all is right at the mechanical level, then factors of disease, nutrition or lethal genes cannot be ruled out.

Erratic hatching?

Unexpectedly early hatching, with thin, weedy chicks, points to too high an incubator temperature. Late hatching results from too low a temperature, but can also be caused by prolonged storage of eggs before incubation. Remedies, as above, involve careful inspection of temperature regulating mechanisms on incubators.

A scattered hatch, spread over a long period, is most likely if eggs of different shapes, sizes or ages are together. Aim for uniformity. Eggs of different strains should be set separately, as should those from different aged flocks. Take into account that older birds tend to lay eggs of larger average size.

Weaklings?

Apart from genetic faults among individuals, whole hatches can be stunted by overheating or lack of humidity during incubation. Small eggs will, of course, hatch small offspring.

'Mushy' chicks are the soft, lethargic birds produced by below average incubator temperatures and high relative humidity, possibly caused by defective ventilation.

Omphalitis, in which the young birds have unhealed navels and a bad smell, is the result of an infection. High strength fumigation of the incubator and meticulous disinfection of all equipment is called for.

Final handling and despatch

Treatment of newly-hatched stock depends a lot on the species and its final destination. Meat strains require the least attention. They may be sexed, if a grower intends to take advantage of the different rates of growth of males and females. Occasionally they will be vaccinated, but quite often they will be immediately placed in boxes, ready for despatch. On the other hand, laying stock, with a longer productive life ahead of them, may receive the 'full treatment', being sexed, debeaked and vaccinated within hours of leaving the shell.

I have described the principle of sexing in Chapter 4. For vent-sexing, teams of trained sexers visit commercial hatcheries on hatch days. They are provided with tables surfaced in non-absorbent material, adjustable chairs, good eye-level lighting and proper washing facilities in the sexing room. They must, of course, wear overalls and footwear supplied by the hatchery, in the same way as other staff, but mostly they are itinerant workers, paid to provide a fast, accurate service before moving on to the next point of call. A fully skilled sexer, after several years of training and experience, will separate up to 1,500 chicks an hour with around 98% accuracy.

Cockerels are generally gassed with carbon dioxide, since male hybrid layers are too light and slow growing to compete with commercial broilers, but in Britain they are also supplied to the poussin trade, which needs low cost day-olds and a quick turnover to make its profits. Those that are killed are frequently supplied to mink farms or zoos, or processed into meal for feedstuffs.

When debeaking is considered necessary, it is done with a special debeaking machine employing a hot blade, which both trims the beak and cauterizes the cut. About a third of the upper and lower mandible is removed and great care must be taken not to touch the nostril and interfere with the bird's breathing. This operation can be carried out at the hatchery or later during rearing, at six to nine days of age, and its advocates point out that it reduces feather-picking and cannibalism, while also stopping food wastage by flicking.

Returnable plastic boxes save packing costs and are used under some circumstances for chicks, but they must be very efficiently cleaned between batches, and many companies still favour cardboard boxes, discarded after a single trip. Commonly supplied in 25, 50 and 100-bird sizes, these can be stored flat and made up shortly before they are required. A good box provides adequate ventilation holes and is reinforced at the corners to allow for stacking. It is usually divided into compartments, to allow the chicks to travel without huddling or crushing. Wood-wool or similar material is also placed in the box for protection.

To allow for early losses it is normal practice for commercial hatcheries to add about 2% to the number of birds ordered, but given modern handling methods, with air-conditioned vans and the use of aircraft for the long hauls, survival rates are often 100%—barring accidents and delays—and the poultryman can rear the extras as a bonus.

6 Brooding and rearing

Being your own poultry expert doesn't necessarily mean doing everything yourself, including growing your own replacements. Many egg producers are perfectly happy to buy point of lay pullets and leave the technicalities of brooding and rearing to others. Still, the idea of starting from scratch is attractive, which may be why all the best success stories start with a clutch of hatching eggs or a handful of chicks bought in a market, a few of which survive to be the foundation of a thriving business.

Forget the romance. That is not the way to start even a domestic flock. If it ever happened it was thanks only to remarkable luck and a lot of determination. Rule one for successful brooding and rearing is to reduce the element of luck by buying good stock. Look for reputable suppliers prepared to offer some guarantee of the quality of their birds. Given a sound start, raising stock from day-old need not be specially difficult or expensive.

When starting in poultry it is tempting to compare the prices of day-olds with reared pullets at eighteen to twenty weeks—or the equivalent part-grown turkeys, ducks or geese—and jump to the conclusion that home-produced birds will be cheaper. Indeed, a domestic poultryman not costing his labour and not vaccinating might just make the sums come out in his favour, but even he would have to account for the cost of equipment, warmth—in the shape of food for broody hens or fuel for the brooder—feed for the growers, shelter and possible losses. Remember that experience can count for a lot in the early stages, and high mortality will quickly cancel out the advantage of not having to pay a rearer his profit margin.

Aside from the financial aspect, there are other incentives for going into brooding and rearing. One for the amateur poultry-keeper is that it is an extension and challenge of his skills. All rearers start somewhere, and some of the best began as enthusiastic amateurs. A more tangible reason for raising your own stock is that it gives you a precise knowledge of their history, and this can be quite important, especially in the case of layers. The lighting pattern they experience as they approach sexual maturity, for instance, has a crucial bearing on subsequent laying performance, and if an egg producer inadvertently sets his pullets back by giving them a shorter day-length than they were receiving in rearing, he will retard lay and cut the overall production of his flock.

The home rearer has a say in whether his growing birds are run on range at all, whether they are raised in cages or on the floor, and what kind of feed they are given. He can monitor weight-gain and start them in lay at just the weight he wants them, with reference to breeder's recommendations. He is in the driving seat.

There is the additional point that birds reared within a relatively short distance of their laying quarters do not have to suffer the unsettling process of a long journey by road, rail or air. Even if you are buying from a rearer this is a point worth keeping in mind. While every effort is made to see that young birds travel comfortably, reared pullets are at a much more sensitive stage of life than day-olds and prolonged journeys are a potential stress. So, other things being equal, buy reared pullets from the nearest local source.

Natural brooding

For small batches of day-olds it is not essential to have any elaborate equipment. I have a friend who has just completed a successful season with five ducks and a drake which he brooded from the start in an open box near a radiator. This is feasible where you have a room with a constant background temperature—at night-time as well—of around 20° C (68° F), with a higher temperature near the birds, and protection from draughts, but I do not recommend it. What it demonstrates is that good stock, given regular attention—and these were family pets—will survive under less than optimum conditions. I hate to think of the mortality if it were attempted with large numbers.

An established poultryman with a limited requirement for replacements is quite likely to turn to the broody hen for aid. And why not? Given adequate and properly balanced rations, fresh water and housing secure against predators and vermin, the mothering bird will take care of the rest, keeping the young warm and encouraging them to feed.

If the broody is the natural mother, simply carrying on her function after incubating the eggs, there should be no complications, but quite often, of course, she is expected to foster others. Broody hens are well known for their willingness to mother not only their own kind, but ducklings, goslings or other young birds. Introducing strangers can be tricky, however. A hen must have spent at least a few days incubating eggs beforehand. She will not expect them to 'hatch' simultaneously, so her new brood must be introduced one at a time, over a period, while a discreet watch is kept to see that she is accepting her new family. Use your discretion about the number she will brood. An average hen will mother about fifteen chicks, but it will depend on her body size. By the same terms she might take ten to fifteen ducklings, which tend to come off heat sooner than chicks, and correspondingly fewer turkey poults or goslings.

Housing can be simple as long as it provides good protection against the weather and keeps out possible marauders—it's surprising how many animals regard young birds as a tasty delicacy—while giving the poultryman easy access to his stock. The triangular shaped fold described in Chapter 5 is useful since it gives the youngsters a corner where they can escape from being trampled by the hen.

A single coop is merely a box made in weatherproof materials, the front having slats set wide enough apart to allow chicks to come and go, while preventing the hen from doing the same. In this way they can be fed the appropriate rearing ration separately from the adult bird. Measuring approximately 51 cm (20 in) square, it has a roof sloping from a height of 51 cm (20 in) at the front to 41 cm (16 in) at the back. This should overlap by 15 cm (6 in) at the front and 7.5 cm (3 in) on the other three sides. It should be possible to close the coop at night, and a removable wooden floor can be fitted.

A development is the double coop, which has two compartments, one for the hen and the other for the chicks alone. This can be 1.2 m (4 ft) long, 53 cm (21 in) wide and with a roof-height again running from 51 cm (20 in) at the front to 41 cm (16 in) at the rear. The smaller compartment, which has a solid door, is 53 cm (21 in) long. The ends and back are made solid. The roof can be made in two parts, and hinged to form lids.

With the double design, the chicks have a refuge from the hen, where they can be fed separately but under cover if the weather is bad. The hen can be fed whole grain, as she would when incubating, while the chicks receive a proprietary starter mash. These simple houses can be placed in pens—ideally about 3 m (10 ft) square for a brood of 15 birds—on short grass which has not been ranged over by other poultry for at least two years.

While ducklings, goslings and turkey poults can be fostered by a hen in these conditions, the fast rate of growth of these species soon makes it difficult for them to use the broody coop. The option is the 1.8 m × 1.2 m (6 ft × 4 ft) house—a garden shed will do duty—with a floor on which straw or wood shavings have been spread as litter. These birds can, of course, be raised by their natural parents, and geese, particularly, are often left to their own devices. Where the foster-mother hen is used she can be removed from her charges after three to four weeks. All will do quite well on proprietary starter feeds, but beware of added medication for chickens, which may disagree with waterfowl. Crumbs of about 3 mm ($\frac{1}{8}$ in) size can be obtained for ducklings. Failing these, broiler crumbs can be used. Similar rations are suitable for goslings, but being natural grazers they appreciate access to good, short grass. Long grass will not do. Turkey poults are well catered for by compound feed manufacturers. Again, crumbs or a coarse mash can be fed.

Each species has its own characteristics, of course, and the water birds are inclined to be much more messy than others, with wet droppings, so that litter will have to be changed frequently if they are housed as suggested above. Similarly, if a grass pen is used, the surround will have to be moved regularly.

The advantages of open-air brooding and rearing are clear enough. The birds derive benefit from sunlight and scavenge a good many of their essential nutrients, minerals, trace elements and vitamins from the insect and vegetable matter they pick up. Pullets that have had to face up to the elements in the growing stages are noticeably tight feathered and spry.

Unfortunately the disadvantages are fairly weighty and equally clear. The main one is that extensive rearing on anything like a commercial scale is a land-hungry operation, and it calls for labour which is ever more costly. Light control can be more difficult. The very fact that birds are free to peck at the soil exposes them to parasitic infections found there. This is why it is inadvisable to run birds over ground recently occupied by other stock which, among other things, could have left coccidiosis to be picked up. For the same reason, long grass, bits of twine or any other oddments can lead to crop impaction.

Unless particular care is taken to treat both coops and mother birds with a safe pesticide the chances of infestation with ectoparasites such as red mite and lice are actually higher than in well-run intensive houses. These can also be inherited, along with other diseases, from visiting wild birds, which are only too happy to drop in for a free feed if a run is not wired over.

A few commercial rearers with a good deal of land do still overcome the difficulties, to supply a specialist service, but generally outdoor rearing is the province of the small-scale operator these days.

Artificial brooding

Since the immediate need of every young bird emerging from the egg is warmth, this is obviously the first consideration of the rearer, who must see to it fairly quickly, in the absence of a hen, if his new stock are not to get chilled, though they will survive reasonably happily in their travelling boxes for an hour or so if their journey hasn't been extended for any reason. This gives a little lee-way for any last minute hitches, but there should not be any. Always have a rehearsal before deliveries arrive, to make sure their equipment and accommodation is prepared well in advance.

Where relatively small numbers of, say, fifty to eighty chicks are to be brooded, it is quite possible to use their own collective body-warmth, without additional artificial heat, by employing a hay-box brooder. This is quite easy to make up. It consists of a box approximately 15 cm (6 in) deep and 91 cm (3 ft) square with a wire floor of 13 mm ($\frac{1}{2}$ in) mesh. A pop-hole 15 cm wide is cut in one side. Within the box a ring of mesh is arranged, with its opening at the pop-hole.

Clean, dry hay—and it must be clean to avoid any risk of the unpleasant fungus infection known as aspergillosis which can arise from mouldy hay—is placed loosely between the walls of the box and the mesh ring. It is also put on the floor within the ring, to a depth of about 5 cm (2 in). The lid of the box is a framework, lined with sacking. It may be reinforced with mesh, to take the weight of the small sacks of hay which are placed on top when the chicks are inside.

This type of brooder can give a satisfactory start for ducklings and goslings as well as chicks, making due allowance for the extra head-room they may need, and their faster rate of growth. A hay-box which may take forty chicks up to eight weeks will accommodate twenty ducklings for perhaps five weeks. Ventilation must be carefully considered and here the loose packing of the hay helps, but the birds should be suffi-ciently concentrated, when all in the centre, to share their warmth. An idea of how closely they will pack can be gained by seeing them in the travelling box, where they rely on the same principle of mutual warmth. As the birds grow the ring of mesh is gradually extended until it is removed altogether.

To give the newcomers a warm reception, also to ensure that the hay is thoroughly dry, hay-boxes can be placed over a convenient heat source for several hours beforehand, but care must be taken to avoid fire.

In operation, the hay-box is quite similar to the broody hen. It can be placed so that it opens into a run 1.8 m (6 ft) long, with boarded sides and a low roof of mesh. Feed and water is then placed in the run and the young birds can emerge to forage for short periods before scurrying back to 'stoke up' again. The trouble is that they do not always find their way back to the warmth unaided and they will need to be herded in and out of the box several times a day for about the first three days. In the meantime keep the pop-hole closed, only opening it for a few minutes at the feeding periods. The time out will depend on the sur-rounding temperature, but must never be long enough for them to show signs of huddling down or crawling under one another for warmth. After the initial stages they can gradually be allowed to come and go as they please. Ducklings and goslings may prove more hardy than chicks, but they still need careful attention, and this management factor is one of the main drawbacks to the hay-box system. Later it is simply necessary to renew the hay litter and see that the birds have sufficient space.

Heating
Many different methods of brooder-heating have been developed over the years, from hot water pipes heated by coal-fired boilers to modern, electronically-controlled catalytic gas burners.

Nowadays electricity is the most widely adopted fuel for small-scale brooding, while gas, either piped or bottled, gets the vote of the majority of professional rearers. Oil also has its supporters, particularly among broiler producers who employ whole-house heating methods. Some-times it fuels central heating systems operating with hot water, and sometimes large heaters which project warm air directly into the house through polythene ducting.

A small batch of goslings is electrically brooded in a hardboard surround

There is something to be said for paraffin or kerosene lamps in emergencies or where alternative fuels are hard to come by. Indeed, there are probably still quite a lot of them about. But they must be regularly refilled and the wicks need to be trimmed to a rounded outline to avoid smokiness, and in general the cleanness and convenience of electricity makes it far more acceptable for domestic purposes.

A single, 250-watt infra-red lamp is all that is needed to heat a group of up to fifty chicks, and since they move away from the heat sooner, it will suit about the same number of ducklings or goslings, despite their faster rate of growth. Turkey poults can also be brooded by this method, but, as can be seen from the table, they need slightly more heat for a longer period.

The brooding area for a small batch of birds can be almost anywhere convenient to the poultryman, its main requirements being that it is out of draughts and—if electricity is being used—has a convenient power point. Once again, a garden shed could be ideal.

Before the birds arrive, set up a circular wall, about 46 cm (18 in) high and 1 m (3 ft) in diameter. This can be made in cardboard or any suitable material you have to hand, its purpose being to prevent the youngsters from straying too far from the vital source of warmth.

Place litter in the ring to a depth of 5 cm (2 in). Various types of litter are obtainable, wood shavings being commonly used. Washed sand is good, or you can take your pick from chopped straw, peat moss, shredded paper or any similar absorbent substance. Whatever flooring material is used must be clean, uncontaminated by chemicals or preservatives, and not previously used as poultry litter.

Now comes a stage of experimentation. Broadly speaking, the operational height for a 25-watt infra-red lamp is a little under 46 cm (18 in) for chick brooding. By placing a thermometer on the floor at several points under the lamp it is possible to see its 'throw' of heat at different heights. Adjust it up or down until the temperature at litter level is from 35°–37.7° C (95°–100° F) over such an area that all the chicks will be able to benefit from it when they are together. Then, outside this radius, the feeders and drinkers can be arranged.

Switch on the brooder several hours in advance of introducing the birds, to warm up the litter and test that the lamp is working properly. Once the birds are in, watch them carefully. Their reaction to heat, for example, is a better guide to optimum conditions than any thermometer reading. If, when they are acclimatized, they crowd in the centre, it is too cool for them. If they scatter in a circle around the lamp, it is too hot. Aim for a fairly even distribution of birds across the brooding area.

After warmth, a chick's first requirement is likely to be water. Food is less immediately important, since the yolk sac, which it absorbs just before hatching, will satisfy its nutritional needs for at least forty-eight hours. This is why young birds can be sent on prolonged journeys, minus food or water, without coming to any harm. But dehydration is a prime cause of early deaths, so water should be waiting for them. It should be plentiful and obvious, but not so deep that they can drown themselves in it, which seems to be a favourite pastime of turkey poults. In very cold weather the chill should just be taken off drinking water.

Sweet jars upturned in shallow dishes make excellent drinkers, adding to the availability of fresh water an attractive, sparkling appearance. Healthy young birds have a natural curiosity which draws them to anything bright or colourful. Sweet jar founts can be improvised or bought, as can other proprietary types of drinker, usually made in brightly coloured plastic.

One drinker may be sufficient for up to fifty birds, but where there is any doubt, err on the side of generosity. If you overdo it you may get wet litter—something to guard against—but at least the birds will not go thirsty. Ducks, in particular, need to be able to immerse their heads— though not their bodies—to help them in cleaning their eyes and clearing their bills of dry feed.

Once your charges are drinking well, which they should be after about two hours, feed can be put down. Fibre flats, known as Keyes trays are frequently recommended as a good method of serving food to day-olds, But any flat container not more than 5 cm (2 in) high and roughly 46 cm × 61 cm (1 ft 6 in × 2 ft), such as a box-lid or a plastic seed-box, can be used. Spread starter mash or chick crumbs on the tray. Some poultrymen scatter cut corn or maize on top as a 'relish' and certainly anything which induces starters—especially slow feeders like turkey poults—to eat, is well worthwhile. If they do not eat you will start to see starve-outs in about three days. Where trouble is encountered, check the lighting. Young birds often seem purblind early in their development and feeders, waterers and brooding area should be well illuminated with at least a 100-watt bulb.

The system described so far, with a single 250-watt infra-red heater, is quite sufficient for fifty chicks, but what if the brood is 100 or more? Obviously it is possible to set up a second hover surround with another lamp, or even use two lamps over a single hover, but there are other options. One is the 'electric hen', a table-shaped brooder on four legs under which up to 100 chicks can congregate. Another, introduced on the European market in 1976, was developed in Canada, and uses the slightly unusual technique of electro-conductive lacquer painted on a polyurethane foam panel to generate heat, rather than the more conventional glowing wire element. The panel, suspended on chains, measures $62.5\,cm \times 42.5\,cm \times 3\,cm$ $(24\frac{5}{8}\,in \times 16\frac{3}{4}\,in \times 1\frac{1}{8}\,in)$ and at $50\,cm$ ($20\,in$) from the ground is said to produce enough heat for 250–300 day-olds.

Both these systems are substantially more expensive than the simple infra-red lamp, but they cost less to run. Claims for the Agrotherm panel just described, produced by the Canespa company, are that its thermostatic control gives it a 25% margin of saving over heaters without a thermostat.

A further option for the poultryman who suspects electricity supplies—and a power-cut does mean a very rapid loss of temperature—is the small version of the infra-red gas brooder which can be run off a camping gas cylinder.

Gradually, as the birds grow, temperature is reduced and space allowance increased, as shown by the guide figures in the table, until the transition is made from brooding to rearing. What form that takes will depend on the species and your intentions, but for ducks, geese and other meat birds it can mean a transfer to grass pens, open-sided sheds or verandahs. Laying stocks have an additional factor, light, to be taken into consideration, as will be seen in the next section.

Intensive systems

Compared with small-scale methods of brooding and rearing, the modern intensive system really is a 'whole new ball-game'. Dealing with large numbers of birds—20,000 or more in one house is not uncommon—a commercial rearer must work to a carefully scheduled plan. He will have a programme of vaccinations to follow, blood samples may have to be taken. Debeaking might be considered necessary. If pullets are being grown a proportion may be weighed at regular intervals. Heat will be phased down, light will be controlled and feed altered as the birds grow. All this can be done by rote, much of it to breeders' recommendations, but the rearer must never stop observing and thinking. Is too much feed being wasted by that feeder design? Does that change in water consumption suggest a disease challenge? Would a change in stocking density for those broilers cure the wet litter problem? (Certainly lighter stocking densities do help to alleviate this sort of situation.)

Intensive rearing presupposes the concentration of a large number of birds under one roof, and this is the way that virtually all broilers, most commercial pullets and turkeys, and to a slightly lesser extent ducks, are grown now. The houses are usually big, well insulated buildings, and, where light patterns are important, they are frequently windowless and environment-controlled.

Although some broilers are raised in cages, floor rearing is generally used for almost all meat birds, while pullets are more evenly divided between floor and battery brooding systems. Other possibilities include the combination of a litter floor with a raised platform of wooden slats or welded wire. Current research is investigating a system of several slatted floors for poultry in a single-storey building, on the logical basis that poultry need less headroom than human operators, and that this should make the best use of available space.

Floor brooding

Most new buildings have a concrete floor, though quite a number of established houses have an earth base, but whatever the foundation it will need a covering of litter. Clean white-wood shavings are widely used commercially, but in recent years they have become scarcer and more expensive, setting off a hunt for suitable alternatives. As a result some producers have installed straw choppers and paper shredding machines to manufacture their own materials. Shredded newspaper is quite good, but avoid colour supplements. Printer's ink on 'glossy' magazines can be harmful to chicks.

Wood shavings may arrive in bulk, in sacks or baled in polythene. One bale weighing 4.8–45 kg (90–100 lb) is sufficient for about 100 chicks. If the house is empty between batches the litter may be placed straight in, but where any delay is involved storage conditions must be dry and weatherproof. Sealed polythene bales are secure enough in the open as long as they are not punctured. The importance of dry litter at the start of brooding cannot be over-emphasized. A day or so before the birds arrive it is spread to a depth of 8–15 cm (3–6 in) across the whole brooding area.

Some effort is worthwhile at this stage to see that the litter is spread evenly, because it is too late once the birds are in. Even a moderately bumpy floor can pose problems if you are only about two inches high. Drinkers and feeders dumped at odd angles are a mark of poor stockmanship, creating possible starve-outs where birds fail to see or reach food or water, while different floor levels under a brooder can cause patchy heating, some areas being warmer than others.

Because day-olds brooded *en masse* do not have the benefit of parental guidance, it is the stockman's duty to become 'mother'. In fact young birds will quickly latch on to any large and dominant figure seen regularly in their early days, providing it moves steadily and gently amongst them. But no human can set them a parental example by pecking at food or water. This must be done by equipping and lighting the 'canteen' attractively and spending some time drawing their attention to it.

As with domestic brooding, Keyes trays or lids often double as auxiliary feeders, but they must be scrupulously clean. Adhering fluff or contamination can be a short-cut to heavy mortality in the brooder house.

More permanent feeders may be of several varieties. Troughs are the simplest, though not the most labour-saving, system for all species. A typical wooden version will be about 12.7 cm (5 in) wide and 6.3 cm ($22\frac{1}{2}$ in) deep, with a lip to prevent undue food wastage and a spinner above to discourage birds from perching on the trough. It can be of any convenient length. Wire mesh of 2.5 cm (1 in) or 2.5 cm × 7.6 cm (1 in × 3 in) can be laid on the feed as an additional precaution against food wastage.

Round feeders, in galvanized metal or plastic, are easily obtained, in different sizes, from several manufacturers. The expensive, but infinitely less laborious alternative for growers is a mechanical feeder, working either on the principle of a travelling chain drawing food along continuous troughing, or of delivery by tube and auger to individual feed pans.

At the outset, adequate space allowance must be made to give every bird easy access to food and water (see p. 83). Position the feeders carefully around, but never directly under brooders, where infra-red rays are liable to 'cook' some of the nutrients, notably those of the vitamin B complex, and create nutritional deficiencies in the young birds. Nutritional programmes for growers can be seen in Chapter 10.

Two main systems of heating are employed in commercial brooding. Whole-house heating may be by hot water pipes in the house, or by warm air projected through ducts from a heater, usually oil-fuelled, adjacent to the brooding area. Spot brooding, on the other hand, in-involves a number of infra-red brooders actually in the house, providing the required warmth at floor-level, but with a lower background warmth in the house as a whole.

Those who use whole-house brooding allow the birds the run of the floor from the start, but in the interests of economy generally hang full length polythene curtaining to restrict the brooding area to a half or a third of the total house. For chickens the curtaining will probably be raised by a foot or so in the third week and completely removed in the fourth week.

Infra-red gas brooders are normally fuelled by 'bottled' liquid propane or LP gas. Some have large, reflective canopies, but the net result is the same, warmth for the birds at floor level. The concise definition given by one manufacturer of a canopy brooder was that it should provide a brooding area of approximately 46.50 sq m (50 sq ft) where a temperature of 36–37° C could be held within $\pm 2°$ C by thermostatic control and with a room temperature 18–20° C (65–70° F) below brooding temperature.

Turkey poults under a
Maywick gas brooder
show good distribution,
neither huddled together
nor spread away from the
heat. Note 'sweet jar'
founts. Wire surround,
to prevent poults from
straying, is often
dispensed with when
brooding chickens

With spot brooding hover surrounds may be put up as with the small-scale system already described; a single commercial brooder may take 500 or more birds, so the surround will be correspondingly large. Turkey poults, in particular, benefit from surrounds to prevent them straying, but broilers are now frequently given the complete run of the floor from day-old. Providing warmth, food and water are in brightly-lit areas to attract them, they thrive.

Battery brooding

Although cages are obviously an added capital expense compared with any floor system, there are several incentives to install them. The most important is the much better use that can be made of the house space. Three times as many birds can be reared in a house with four-tier cages as can be reared in an equivalent house on the floor. Generally all the birds are started in the top tier, heated by gas or electric brooders from above. As the brood come off heat, they are divided into equal proportions and occupy all the tiers.

Apart from the space advantage, it is often argued that pullets conditioned to cages from the start take more readily to the environment of the cage-laying house. Management is simpler, the birds being easily seen and accessible, while the need for litter is done away with. The fact that they are off the floor can also make a difference to disease precautions, since there is no risk of their picking up floor-borne organisms like the oöcysts of coccidiosis.

The idea of cage brooding is not new, but only comparatively recently have the full benefits of the system been exploited. Simple tier brooders in which heat was supplied at one end of the compartment by oil, gas or electric burner and the chicks had the run of the remaining cage area, were in use many years ago. At least one ingenious designer connected the functions of incubators and brooders, arranging cages alongside the incubator and using a common heat source for eggs and chicks. But all the early tier brooders, some of which can still be found in use occasionally, were essentially small-scale, and restricted to the first three or four weeks of the birds' growth. After that they must be moved to separate rearing quarters. Modern cage brooders, on the other hand, are little different from laying batteries. Indeed, with progressive adaptation as the birds grow, many of today's cages will take them right through from day-old to point of lay and even end of lay if necessary.

Adaptability is a particular feature of cages. By adjustments in the size of floor mesh and side wires, birds from the tiny quail and guinea-fowl chicks to some breeds of turkey breeder hens can be accommodated. Using special 'trampoline' plastic floors to provide a more yielding surface, cage designers now claim to have overcome the problems originally encountered with broilers, namely foot troubles and breast blisters, so that these heavier types of chicken can be successfully reared to slaughter ages.

Layers are frequently brooded in batteries. As they grow, chicks are distributed among the middle and lower tiers

Much the same principles apply to cage brooding as to birds on the floor. Prepare the house well in advance. Test the heaters and have them running to bring the temperature up to the required level in the brooding area before the birds arrive. Some rearers use paper, others a fine-mesh plastic overlay for the floor in the first few days. This is laid on the existing wire floor of the cage. When it is removed will depend on the type of bird being reared and its rate of growth.

Glass water founts are often used for the first day or so, as an inducement to drink, but if sufficient attention is paid to management, young birds can be quickly persuaded to take water from nipple drinkers or drinker cups which will, of course, be set at their lowest level. To give them a clue, nipple drinkers should be triggered frequently by the poultryman. Any floor overlay should come right up to the drinker, so that the birds are not discouraged by having to cross an open wire floor. Likewise, feed trays should be where they cannot miss them, but their attention should also be positively drawn to the feed troughs outside the cages, by running a finger, or preferably something shiny, like a metal key, through the feed in the trough. At all times observe the birds. They should be active, alert and inquisitive from the moment they come from their boxes. This happy disposition only alters in the face of disease, an unfavourable environment, or failure to eat or drink. An effective stockman sees that they stay happy.

Lighting

Daylength—the period of light received by birds in each twenty-four hours—is very important in the growing stage. It affects the age of maturity, influencing food consumption, body-weight and, in the case of breeders and layers, the size and number of eggs laid.

The length of lighting, not its actual brightness, is the stimulus, but do not underestimate the sensitivity of the eye. Quite often just the light that filters into a so-called 'dark' house, through ventilators or cracks in the structure, is sufficient to stimulate a reaction in growers and upset the light pattern. Spend a few seconds in a blacked-out house on a sunny day and you will find it quite revealing, in every sense. To operate an effective light control programme, ventilator inlets and outlets must be properly baffled, cracks around doors and windows filled, and any other light 'leaks' sealed. Intensity should not exceed 0.04 foot candles during dark periods. Ministry of Agriculture advisers can check this with special light meters or fieldsmen from breeding or feed companies can be asked to help.

Several methods are used for expressing light intensity, one foot candle being the same as one lumen per square foot. In practical terms, this translates into the amount of light spread by a 50-watt bulb over an area of 10 ft × 10 ft. The metric version is the lux, which is one lumen per square metre of surface.

During light periods in rearing a level of 1 to 5 lux is generally aimed at, and since light varies considerably with distance, readings should be taken in the area furthest from the light source.

All this may suggest that windowless houses are indispensible. Of course they are not. They allow the application of special lighting patterns and close control over the rate of bird development, but natural daylight though needing to be supplemented at certain seasons by artificial light, is perfectly acceptable and has the advantage of coming free.

All artificially brooded birds should be started on bright light for up to twenty-three hours a day to give them every opportunity to familiarize themselves with their surroundings and find food and water. After not more than a week, laying strains—including breeding birds of all species—need to be subjected to a reducing period of light if they are not to be precocious and start laying early, undersized eggs. On range or in windowed houses this will depend on the season.

In the northern hemisphere daylight diminishes naturally from June to December. This means that pullets hatched from April to August will experience a falling light pattern in the crucial stage of their development to twenty weeks. Those hatched early come on to it from about eight weeks of age. At other times of the year, from September to March, after the first week at twenty-three hours light, supplementary lighting

Light intensity at various distances from lamps

	Distance from lamp		House interior surfaces	
			dull	shiny
	metres	ft.	foot candles	
15 watt tungsten filament lamp	0.6	2	2.0	3.3
	0.9	3	0.9	2.0
	1.2	4	0.6	1.4
	1.5	5	0.4	1.1
	1.8	6	0.3	0.8
	2.4	8	0.1	0.8
25 watt tungsten filament lamp	0.9	3	1.7	4.0
	1.2	4	0.9	2.9
	1.5	5	0.6	2.1
	1.8	6	0.4	1.5
	2.4	8	0.2	1.1
40 watt tungsten filament lamp	0.9	3	3.3	>5
	1.2	4	1.9	>5
	1.5	5	1.2	3.8
	1.8	6	0.8	3.1
	2.4	8	0.5	2.5
60 watt tungsten filament lamp	1.2	4	3.2	>5
	1.5	5	2.0	>5
	1.8	6	1.4	4.3
	2.1	7	1.0	3.8
	2.4	8	0.7	3.6
Warm White fluorescent tubes	1.2	4	10.0	>11
	1.5	5	6.5	>11
	1.8	6	4.5	8.7
	2.1	7	3.3	6.4
	2.4	8	0.7	4.9

Type of shade		Cage Tier	K for positions opposite lamp	K for positions between lamps
Plastic cone		Top	1.1	0.95
		Middle	1.15	1.1
		Bottom	1.25	1.15
Plastic bell		Top	0.7	0.7
		Middle	1.0	0.7
		Bottom	1.05	0.9
Plastic bulb (2cm band around equator of lamp)		Top	0.2	0.9
		Middle	0.7	0.75
		Bottom	0.7	0.75
Painted bulb (2.5cm spot opposite nearest cage)		Top	0.2	0.95
		Middle	0.7	0.1
		Bottom	1.0	1.0

Effects of lamp shade designs

This system, worked out by Ministry of Agriculture advisers, allows the amount of light reaching birds in cages to be calculated. Light points are usually at 3 m (10 ft) centres. To find the effect of the shade, multiply the K factors quoted by the light intensity of the chosen light source in the table. E.g., birds 0.6 m (2 ft) from a 15 watt bulb in a plastic cone will receive $1.1 \times 2.0 = 2.2$ ft candles if they are in the top tier in a house with dull walls, or $1.1 \times 3.3 = 3.63$ ft candles in shiny surroundings.

is continued but stepped down by a set amount each week. Thus, at British latitudes, daylength is cut by thirty-five minutes a week for pullet chicks hatched in September, thirty minutes a week for those hatched in November and twenty minutes a week for any hatched in December to March. By the time they reach twenty weeks of age all such birds will be receiving natural daylength. No particular rule determines whether light adjustments should be made at the beginning or end of the day and the poultryman can fit this in wherever most convenient with feeding or other schedules.

What applies to pullets applies to a varying extent to other species, depending on the speed with which they normally reach maturity. Turkey breeder hens, for instance, do not mature sufficiently to produce good-sized, hatchable eggs until about twenty weeks for light breeds or thirty weeks for heavy breeds. Natural lighting for them, therefore, extends from April into September. When light control is practised at other times they need a reducing pattern at least from sixteen weeks onwards (after the first four days reduced to a sixteen hour day).

Where windowless houses are used, seasonal changes in daylight can be forgotten, of course. Here, providing the golden rule of constantly diminishing light for growers and constantly increasing light for layers is observed, any number of variations can be applied. Indeed, every breeder seems to recommend a different light pattern to give his bird an advantage, sometimes real, sometimes imagined, in final production, but in general the systems fall into two groups; those that give a constant daylength and those involving a step-down sequence.

Arguments for the constant daylength system are that it is simple to operate—after the first brooding days, light is simply reduced to six, eight or ten hours and held there until twenty weeks of age—the birds mature earlier and lay more eggs in the season. Supporters of the step-

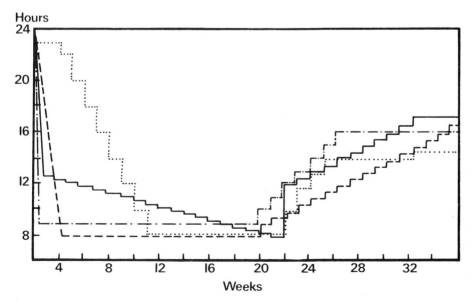

Light patterns
This composite graph shows various lighting periods recommended by different breeders, some on constant day, some on 'step-down' principles. The crucial factor is that lighting periods decrease during rearing and increase when birds come into lay at eighteen to twenty weeks of age. Some programmes include a further increase, by $\frac{1}{2}$ hour a week from thirty-six weeks to a maximum of seventeen hours.

82 COMPLETE HANDBOOK OF POULTRY-KEEPING

down system, on the other hand, find that it tends to delay maturity and produce more large eggs from the start of lay. Steps may be steep drops of two hours a week, bringing the birds to an eight-hour day by nine weeks, or a series of shallow reductions bringing them to the same level by twenty weeks, and on the basis that breeders know the characteristics of their birds best, it is advisable to adhere to their recommendations.

Under controlled environment conditions, broilers as well as growing pullets are subjected to special light patterns, but for quite different reasons. Here the object is to encourage feeding and stimulate fast growth. Again, there is wide scope for theorizing. Many broiler producers have adopted a policy of continuous light, to allow the birds maximum opportunity to find the feeders. Better results have been reported with the use of intermittent light, however, possibly because alternating short periods of darkness with short periods of light enforces a rest after feeding activity and improves the efficiency of digestion. At Plumpton Agricultural College in Sussex, England, a programme of twenty-three and a half hours light a day for the first ten days, eighteen hours from ten to twenty-one days, and then two hours light and two hours dark until slaughter resulted in an advantage of 250 g (10 oz) in average liveweight over birds reared on continuous light and 160 g (5¾ oz) over those on natural daylength in a windowed house. A trial in the United States involving 250,000 birds was reported in *Poultry World* to have led to savings of $12,000 using programmes of 2L : 4D and 1L : 2D and such results seems to be generally repeatable. From McGill University, Canada, R. B. Buckland concluded that the best programme for improved performance was 1L : 3D.

High light intensity is often blamed for outbreaks of vent-packing and cannibalism among birds. Certainly brightness and bullying do tend to go together, but I would be very careful about drawing hasty conclusions here. Light colour could be more important. Many egg producers have found, for instance, that fluorescent tubes which give off more light in the blue-green range of the spectrum than equivalent tungsten lamps can be used at much greater intensities without trouble. Other possible factors contributing to what is popularly called vice are tight stocking densities, other stressful conditions or just plain boredom when feed troughs have been closed.

Where every other avenue has been checked, if vice persists with layers—and sometimes turkeys—there is probably nothing for it but to debeak the birds. Some hatcheries do this as a matter of course at day-old. In other cases it is a job for the rearer, usually when the chicks are six to nine days of age, though it can be done at other ages. Sometimes ducks are prone to feather pulling and farmers resort to beak trimming, either with scissors or a debeaking machine.

Brooding temperatures

	Days 1–2	Days 3–7	Week 2	Week 3	Week 4
CHICKENS	35–37.7° C	35	32.2	29.4	23.9–26.6
	95–100° F	95	90	85	75–80
	(whole house start 33° C then 29° C after 1st 12 hours)				
POULTS	37.7° C	36	32.2	28.3	24.4
	100° F	97	90	83	76
DUCKLINGS	29–32° C	26–29	23–26		
	84.2–89.6° F	78.8–84.2	73.4–78.8		
GOSLINGS	32° C	29	22		
	89.6° F	84.2	71.6		
QUAIL	35–37.7° C	35	32.2	29.4	23.9–26.6
	95–100° F	95	90	85	75–80
GUINEA-FOWL (KEETS)	35–37.7° C	35	35	32.2	29.4
	95–100° F	95	95	90	85

Space allowances

		Age in weeks	Floor space sq ft/bird	Sq m/bird
CHICKENS	Litter	0–4	0.5	0.04
		4–8	1.0	0.09
		8–16	1.25	0.12
	Cages	0–5	Up to 0.17	0.01
		5–9	,, ,, 0.34	0.03
		9–16	,, ,, 0.50	0.04
TURKEYS	Litter	0–4	0.5	0.04
		4–6	0.75	
		6–8	1.0	0.09
DUCKS (fatteners, groups 40–50)		1–2	1.0	0.09
		2–3	1.5	0.13
		3–4	2.0	0.19
		4–5	2.5	0.23
		6–8	4.0	0.37

7 Housing and management

Broadly speaking, poultry husbandry systems fall into three main categories: those that are suitable for home producers; those which fit into a farm situation, where poultry may be run as an adjunct to other activities; and those which are purely suited to specialist commercial poultry keepers.

If you have only a garden, of any size up to an acre, your interest will plainly be limited to domestic production, though that can still involve a good number of birds. Apart from the obvious domestic hen, an ordinary garden has room for practically any of the poultry species from quail, bantams and guinea-fowl to ducks and even turkeys in small groups. Where the land extends to more than an acre, with a fair area of grass, geese come into the reckoning, though with their notoriously loose, messy droppings they will not be popular on lawns. Really, geese are probably best adapted to farm surroundings or the special facilities of the professional producer.

The amount of time and money you can devote to poultry-keeping naturally has a direct bearing on the style of housing and management you adopt. A full-time poultryman wants facilities which enable him to keep the maximum number of birds efficiently for his labour. In today's competitive world that inevitably means some form of intensive production. The hobbyist, on the other hand, with no such pressures, can afford to use more traditional methods. Although it has been superseded for commercial production, outdoor management can apply to both domestic and general farm operations.

The first consideration with any outdoor flock is to prevent them from straying, while protecting them as far as possible from marauders. An enclosure of wire or plastic mesh, 1.5 m–1.8 m (5–6 ft) high will be sufficient to contain any flock, although it may be necessary to clip the wings of the more free-flying species, including guinea-fowl, ducks and geese. As a guide, timber posts set 3 m (10 ft) apart, need to be about 5 cm (2 in) square. Corners may be more substantial. Treat all wood with preservative, particularly where it will be at or below ground level—observing any instructions about toxicity to animals. The main rule for metal or other fence supports is that they should withstand outdoor conditions. Sink the posts about 30 cm–45 cm (12–18 in) into the ground. Netting, for fowl-runs, is generally about 5 cm (2 in) mesh, 18 gauge. For young birds or smaller species, 2.5 cm (1 in) mesh may be used. Both to stop birds working their way under the fence and to prevent raiders from burrowing in, the mesh is often let into the soil for a depth of several inches, but if the enclosure is of any size the necessary trenching will be quite laborious and many poultry keepers settle for turning out the bottom foot or so of net and staking it down securely. At the top it can be left floppy, to deter climbing animals, although no fence should be regarded as more than a first line of defence against a hungry predator. Gates need to be carefully considered when making an enclosure. Not only should they hang properly and have some simple 'no-hands' closing device like a spring or a weight on a string, but they should also be wide enough to take any necessary implements. In an orchard, for instance, access might be needed for tractors, carts or mowers, as well as feed and equipment for the birds.

Wing-clipping, if carried out correctly, is painless for the birds. It is applied to adults with fully-developed plumage and consists of cutting back the 'primaries'—the main outer feathers, separated by one short axial feather from the 'secondaries'—on the wing only. This trimming of the flights to about half their length on just one side, which can be done with a good pair of kitchen scissors, upsets the bird's flying balance and deters it from gaining height.

Pinioning, which has the same effect, is a more radical operation, performed on young birds, usually in their first ten days. For this a sharp, sterile knife should be used. A single wing is placed on a flat cutting surface and its tip, as close as possible to, but not including the 'thumb', is removed. In this case the flight feathers are stopped at source and providing it is done properly, the bird is permanently debarred from flying.

What applies to the enclosure in an orchard, or on an acre or so of open ground is equally valid, and in some ways more so, for a small run. Land on which any species is held for a period unavoidably receives the droppings and builds up a residue of potential infection, notably of parasitic diseases such as coccidiosis, whose oöcysts live quite happily on the soil to await each successive generation. If the stocking rate is low and the area relatively large, the build-up might be almost unnoticeably slow, but it will exist. With a smaller run the concentration is that much greater, coupled with the fact that scratching hens, the birds most usually confined in this way, quickly reduce any grass to a bare mud-patch. It is best, therefore, if at all possible, to have an alternative site and to change from one to the other at least whenever the flock is changed. Young birds, incidentally, should never be added to an old flock piecemeal. They are highly likely to suffer cross-infection from the older birds. To avoid this you should adopt a policy of 'all in', 'all out'.

Now for housing. Do not imagine you are doing birds any favours by providing them with a house, however well designed for them. Remember every species originated in the wild and either roosted in trees or, in the case of waterfowl, sheltered among reeds and vegetation during the night. All have an excellent weatherproof covering of feathers, and this arrangement worked well for several million years before man came on the scene.

Poultry houses were introduced as a convenient management aid for humans. With that in mind, it is surprising how inconvenient, from a stockman's point of view, many of them are. When designing a house or selecting one 'off the peg', go for functional simplicity. Make life as uncomplicated as possible for yourself as well as for your stock. Consider: is it going to be easy to feed the birds? Can they be readily seen and reached if necessary? Is proper provision made for manure removal and will it be simple? If they are layers, can eggs be collected without bother? Are there dirt traps which could make a thorough clean-down difficult? All these questions, you will note, centre on convenience for the operator. It pays to be at least that selfish at the start. There are other matters, too, such as proper weather-proofing of materials and price, to take into account. But although this may seem to put bird comfort a long way down the list it should not be ignored.

Ironically, it is precisely when birds are under cover that their troubles start. They may be safely shut away from animal raiders, but a closed house does not offer much scope for movement if conditions become excessively cold and damp, or hot and airless. Nor is there any escape from mites and similar pests. The more elaborate the system, the more care is needed to see that it functions properly. A power failure in a windowless, controlled environment house can be disastrous, with rapidly climbing temperatures and a suffocating atmosphere. The next chapter looks at the specialized requirements of environmental control.

With smaller systems of housing, in the moderate British climate, try to site units facing south, south-east or south-west, and backing cold northerly winds. If ventilation, via slatted or netted openings or windows, is adequate, condensation will not be a problem. Pests are dealt with by applying the hygiene precautions detailed in Chapter 11.

Domestic systems

'How many birds should I keep?' This is the perennial question among beginners and it is, of course, fundamental to the type of management adopted. Since nearly all domestic poultry-keepers think automatically in terms of standard layers, it really comes down to 'how many eggs do I want?' This will depend on whether the aim is simply self-sufficiency in eggs for the family, or extends to selling a few dozen a week at the front gate.

An efficiently kept modern hybrid will lay more—occasionally much more—than 240 eggs in her first season of approximately a year, an average of four to five per week, so four such layers will provide upwards of sixteen to twenty eggs per week. Obviously family needs vary, but such an output is around the average requirement for a family of four. You may prefer to allow for disease losses or production falls by keeping six birds, in which case any surplus makes a welcome present for friends or relatives.

Equipment manufacturers have been quick to recognize and cater for this end of the market and come up with special compact designs and some complete package deals—house, birds and a supply of feed— in poultry and farming journals. The following example demonstrates the ingenuity of some of these systems. This accommodates six birds on a ground area of 97.5 cm × 1.05 m (3 ft 3 in × 3 ft 6 in) by using a two-storey arrangement, 91 cm (3 ft) high. It can be stood on any suitable piece of ground and the hens use the lower section as a run. The upper compartment has a slatted floor and is fitted with a nest box, food and drinking troughs. It is reached by a ladder which is raised at night to shut the birds in. Light enough to be lifted by two people, even with the birds inside, the unit can be easily shifted when the run conditions demand it, and in fact a section of wire is detachable, to allow the birds into a larger wired area, if desired.

The walls and roof of this professionally made structure are of plastic-coated metal sheeting, with a pressure treated timber frame. Although I have no personal experience of the system and could envisage some possible drawbacks in the rather cramped positioning of water and food in the roosting area, it does feature those elements of functional simplicity referred to earlier.

Everything, including point of lay pullets and a 25 kg ($\frac{1}{2}$ cwt) bag of feed—equivalent to about a month's ration—can be supplied in this deal, to give the novice an easy start in home farming for less than half the price of an average freezer. A number of similar package-deal options have been developed and can be found advertised from time to time in the farm and poultry press and local papers.

Even simpler than the free-standing unit is the miniature battery, which can be set up in a garden shed or similar dry, well-ventilated building. Hens are kept in cages and managed in much the same way as those in large-scale commercial units. This system is worthy of the domestic poultry-keeper's consideration, making the birds accessible and easy to attend and keep clean; a high standard of management can give excellent results.

Beyond what might be called the first stage domestic system, the poultry-keeper is faced with the alternative of buying or building a more extensive unit. For layers this may take the form of a conventional house and run; a verandah, in which the birds run out on wire; an apex fold unit; an ark, or any number of variations.

Buying a new house saves time and trouble and carries with it the benefit of a manufacturer's experience in building and supplying large numbers of similar units for poultrymen. Its obvious drawback is the high capital cost, but this needs to be measured in 'per bird' terms. If a house will hold more birds, or last longer than, say, a second-hand alternative, then in cost per bird it could be value for money. The same will be true if bird productivity is better.

Second-hand units, on the other hand, have given many poultrymen a good start in the business, at relatively low cost. They need to be checked carefully, particularly if they are timber houses, for signs of rot or breaks in the weather-proofing. Beware of parasites like red mite, which collect in woodwork cracks during the day and emerge at night to plague the birds. And on no account put birds into any second-hand structure until it has been thoroughly cleaned down and disinfected.

Money, if not time, can be saved by building one's own units and it is not a particularly difficult job for the do-it-yourself enthusiast if he can pick his way through the tremendous choice of designs and materials. Brick, concrete block, corrugated iron, aluminium, asbestos cement, straw and plastics have all been used with varying degrees of success, although timber remains the general favourite.

Despite the wide range of shapes and sizes, domestic poultry systems must have certain features and one of these is the amount of space they give to the birds. With adult hybrid layers this usually amounts to approximately 1.15 sq m (1.25 sq ft) of floor area per bird, so a conventional style house of 2 m × 1.20 m (6 ft × 4 ft) will accommodate eighteen birds and one of 1.8 m × 2.40 m (6 ft × 8 ft) will hold forty. Roughly twice as many growers up to eight weeks or bantams may be held, but the final criterion is always how comfortable the birds look in the house. While birds congregate naturally and enjoy mutual warmth, the poultryman who deliberately overcrowds them for the sake of increased productivity will find it has precisely the opposite effect.

It is quite possible to accommodate layers or growing stock in a simple, box-like structure no more than 91 cm (3 ft) high, with a pophole at one side allowing them access to a run which can be roughly twice the area of the 'coop'. Since the fences of the run will also, logically, be 91 cm (3 ft) high, it will need to be roofed with wire.

The box is raised on short legs just sufficient to allow free passage of air beneath and floored either with 2.5 cm × 7.5 cm (1 in × 3 in) welded mesh or wooden slats. Where wire mesh is the flooring, the birds can be furnished with perches or a slatted platform to raise them a foot or so higher above ground level and encourage their sense of security. To allow them to be reached, the roof, or part of it, forms a detachable lid. It should be covered with felt or similar waterproofing, and angled to carry away water. Holes or slots near the top of the box wall promote air-flow, but must be arranged so that moisture does not penetrate. It is convenient to have external nest boxes (discussed on page 92) for this type of system, although with apex houses, which are related to it, the nests normally fit inside.

These small units are invariably mobile. Droppings fall through the floor, so saving one of the worst cleaning-out chores, and every so often the house can be lifted to another site. This theory breaks down if the amount of land is limited, however. Then a droppings board becomes necessary under the floor, from which the manure must be removed every few days.

Here the argument strengthens for a more permanent structure which will accommodate twenty or more birds under one roof without difficulty. This is the ubiquitous back-garden house. Frequently it has just one run which is forked over between batches, but if at all possible, two separate runs should be provided, so that one can be rested while the other is in use. It only involves another pophole on the house itself. Try to arrange the runs to derive the maximum benefit from available daylight while being protected from weather extremes.

Normally, this type of building stands 1.2 m–1.5 m (4–5 ft) high. It has a door to give the stockman access to the birds, and one or more wire-mesh windows, equipped with shutters which can be closed across them at night. Light in a house with a run is really only important to the birds during the day only for finding their way to nest boxes, but in many houses the windows are at least as important in ensuring adequate air circulation. In addition, movable louvre boards can be fitted into the highest part at each end.

Earth floors are quite common in small-scale poultry houses, but if used they should stand 7.5–10 cm (3–4 in) above the surrounding ground. The site should be dug out to a depth of 15 cm (6 in) and filled with clinker, broken bricks and similar hard-core to give it good drainage. The disadvantage of the simple earth floor is that it is almost impossible to tamp the earth in hard enough to prevent it being worked into holes by the scratching birds. If it is used it should be overlaid with a litter of woodshavings, straw or peat moss.

Solid wooden floors add to building costs but bring several benefits. The boards should never be less than 19 mm ($\frac{3}{4}$ in) thick and they should be tongued and grooved. Joists of 5×5 cm (2×2 in), arranged at not more than 61 cm (2 ft) centres are sufficient for a 1.8 m $\times 1.2$ m (6 ft $\times$ 4 ft) house. With the solid floor, the house can be raised on brick or concrete piers at each corner to a height of about 30 cm (1 ft). This allows air to circulate under the floor helping to keep it dry and free from rot; it deters rats who love to nest in enclosed spaces under floors; and if part of it is accessible from the run, it provides a dry area for the birds to make a dust bath. The piers must be strongly based, however, and checked periodically to see that they do not subside and cause the house to tilt. A downward facing metal cowl round each pier will prevent raiding animals from climbing up.

The most permanent flooring, and the most acceptable for larger houses, is concrete. The site should be prepared on a hard-core base similar to that described above and carefully levelled. On this a 'raft' of concrete is laid, about 7.5 cm (3 in) deep, which can either be bought ready-mixed or made up on the spot, using 1 part cement and 6 or 7 parts of gravel and fine sand. If you go for the more expensive ready mix, remember it sets quite quickly and have some friends standing by to help with the heavy work of spreading. Put the first half directly on the hard-core, then lay a damp course—strong polythene is very suitable—before topping off with the remaining concrete. It should project above the surrounding land, so wooden shuttering will be required to retain it while it sets.

Walls for timber houses are generally made up in tongued and grooved match-boarding which is seen as vertical planking, or weather-boarding, where the overlapping planks are arranged horizontally. Boards should not be less than 15 mm ($\frac{5}{8}$ in) thick.

The positioning and design of pop-holes need to be considered at this stage. There may be one or two, depending on whether the house gives access to different runs at different times. A hole of 25 cm × 25 cm (10 in × 10 in) is sufficient for any size of layer. It should be placed about 7.5 cm (3 in) above interior floor level so that litter is not trailed out of the house. At night, the pop-hole must be closed and the easiest way to do this is with a vertical, sliding shutter. Various ingenious designs have been developed for closing the aperture, some even allowing the birds to open it themselves by applying weight from the inside, but the simplest method is a rope or wire and pulley which enables the poultry-man to raise and lower the slide without struggling into the house or run. The shutter can be placed outside or inside the house, bearing in mind that if there is room inside, it is less apt to get weathered and stick in the runners.

Still pursuing the need for simplicity, the roof on virtually any domestic poultry house, certainly up to a span of 3 m (10 ft) can be of the single pitch, lean-to type. Give it a slope of not less than 1 : 6. In other words, a house that is 1.8 m (6 ft) in depth and has a front height of 1.8 m (6 ft), should be 1.5 m (5 ft) high at the back, to give a drop of 30 cm in the 1.8 m (1 ft in the 6 ft) span. This applies to a roof of timber—which should be substantial, preferably 19 mm ($\frac{3}{4}$ in) planking—overlaid with roofing felt. For other materials, such as corrugated iron, the pitch may need to be greater. Commercial scale houses, which regularly span 12.20 m (40 ft) or more, almost invariably have gable roofs with a centre ridge which keeps down the overall height while allowing the most convenient arrangement for ventilation and services in the building.

A lot of people believe that fowls take naturally to perches, but this is a fallacy. They will roost just about anywhere possible, above floor level, their main instinct being to reach the highest convenient point away from their natural enemies on the ground. Perches should therefore be above the level of the nest boxes and either at a common height, or very slightly stepped up, with the perch nearest the back being the highest, to encourage orderly roosting, since that will be the first one occupied.

Modern layers can be allocated 18 cm (7 in) of perch space each—two 1.8 m (6 ft) perches for eighteen birds in a 1.8 m × 1.2 m (6 ft × 4 ft) house being more than ample—but up to 25 cm (10 in) may be allowed for turkeys. Use smooth 5 cm × 5 cm (2 in × 2 in) timber and round off the top edges. Fix slots or supports so that the perch nearest the wall

This henpen by Park Lines & Co. holds 8-10 layers. Note nest box, moveable tray for droppings and run area extending below house

is not closer than 2.5 cm (1 ft) to it and there is 38 cm (15 in) between perches. They need to be removable for cleaning, and dipping the ends into paraffin or kerosene once a week will destroy any mites that may be present.

It is usual to have a droppings board about 15 cm (6 in) below the perches and 46 cm (18 in) above the floor so that the birds have the use of the floor without stepping in their own muck. They void about half their droppings when perched. The board can be of any reasonably non-absorbent material, supported on a simple framework, asbestos and glass occasionally being used in preference to wood, although greater precautions must obviously be taken against breakage. Droppings are highly corrosive to many metals, however. Again, the board should be removable.

About 46 cm (18 in) width of board should be allowed beneath each perch, so a single board under two perches will be 1 m (3 ft) wide. How often it needs to be cleaned should be judged by the speed at which droppings build up, and their condition. Wet muck causes a quick deterioration in house environment and may need to be removed daily but relatively dry droppings can be left to accumulate for a time, provided that due attention is paid to control of flies (see Chapter 11) and the need to keep the atmosphere well ventilated and free of ammonia and unpleasant smells.

Slats are an alternative to perches as a roosting area. They use more timber—5 cm × 5 cm (2 in × 2 in) battens at 5 cm (2 in) spacing—but if made up in the form of a platform over a droppings board, offer the advantage of keeping the birds away from the board, while providing more roosting space per individual. To ensure that the manure falls through without clogging, the slats should be chamfered down to a narrower width at their lower side.

External nest boxes, reached by the birds through a hole in the wall, have several points to commend them. They leave more space for the birds inside the house, while the lid, which also serves as weather-protection, on the outside, makes it easy to collect eggs without disturbing the flock. They can also be easily inspected and cleaned out when the litter is changed.

Opinions differ on nest dimensions, but you can reckon on 38 cm (15 in) wide, 30.5 cm (12 in) deep and 40 cm (16 in) high at the entrance for ordinary layers. One nest for every five to six birds is normal, so three will be needed in a house for eighteen birds. At the entrance there should be an alighting platform 10 cm (4 in) wide, and a vertical lip of about 7.6 cm (3 in) to prevent the nest material—traditionally straw, although wood-wool is also suitable—from being strewn about.

Structurally, the nest can be all wood, using, for instance, 6 mm ($\frac{1}{4}$ in) exterior grade plywood or it can have a fine wire-mesh base, to help ventilation. If it is outside, the lid will slope, and be covered with roofing felt, to stand up to the weather. Interior nests are also given a sloping cover but here it is at an angle of $45°$ to discourage perching. Occasionally nests are arranged in tiers, or the divisions between a bank of nests may be left out, to make one long communal nest. Nest boxes should be positioned so that the entrance is sheltered from direct light. This appeals to the hen's instinct for hiding her eggs, and encourages her to use the nest rather than laying on the floor. Relative darkness also reduces the chance of inquisitive pecking by one bird at the temporarily exposed oviduct of another which has just laid, that can develop into the habit of vent-pecking. Raise nests about the eye level of birds on the floor but if possible not higher than the perches. Where this is difficult to avoid, as in the case of tiered nests, a front flap should be provided, to shut the nest at night and prevent birds from roosting in it.

Egg eating, sometimes a troublesome problem among layers, can be countered by sloping the nest floor so that the eggs roll away to a tray or compartment out of reach of the birds.

Apart from a litter of woodshavings, straw or peat moss may be more valuable on earth or concrete floors than on wood, no other 'furniture' is needed in the simple unit with a run. Feed and water can be supplied in equipment outside the house. This can be as basic as you like, depending on how much management you want to undertake. For instance there is nothing wrong with a shallow open pan or trough for water—except that it may need filling more than once a day, is apt to get stepped in, tipped up and frozen over in winter. To save these worries, a properly designed drinker is well worthwhile, especially in view of the extreme importance of constant fresh water to the birds.

Weather and the state of production in a laying flock both have a bearing on water consumption. Layers in full production require about twice as much as those at the commencement of lay, and at peak requirement a hundred birds will be getting through 18-22 litres (4-5 gal) daily, an average of 200 ml ($\frac{1}{3}$ pt) each.

Mobile fold unit

The drinker should, ideally, have sufficient reservoir capacity to meet the needs of all the birds in the flock for at least a day, and if the water level can be seen at a glance, so much the better. The drinking space should be adequate, but not so wide that every sort of dirt and contamination seems attracted to it, and the drinker should be stable enough not to be easily knocked over. It needs to be sited where it has reasonable protection against freezing, but is easily accessible for cleaning. In fact, birds do not like very cold water and in winter it is a good plan to maintain their intake by warming it to around 13° C (55° F).

There are various drinkers on the market which fulfil all the main requirements, or you can rig a home-made system. Some poultry-keepers avoid the risk of forgetting the daily refill by plumbing mains water into their unit. This is done via a ball-valve feeding into a plastic cistern, which must be covered and away from weather extremes, and is the procedure used by most commercial producers. Water is piped from the cistern to a trough or container, or a system of drinker nipples let into the tube.

As with drinkers, so feeders can be quite elementary, just a pan being sufficient for a small number of layers, providing it allows some 10 cm (4 in) of 'elbow room' for every bird. But again, a few refinements can ease the chore and improve the feeding efficiency of the flock.

If a trough-shaped feeder is employed it will need to be 1 m (3 ft) long for eighteen pullets. For wet mash it can be made in galvanized metal or plastic but wooden troughs are widely used for dry mash. In cross-section it should be straight-sided, about 20 cm (8 in) wide and

12.7 cm (5 in) deep, with a 2.5 cm (1 in) lip at the top to save costly food from being scooped out and scattered. Various devices are used to prevent birds from getting into their feed or flicking it about, one of the commonest being to place a piece of welded wire mesh directly on the feed. Often a spinner is placed over the trough about 5 cm (2 in) above the level of the sides. This is usually just a batten, about 8 cm (3 in) square, with a nail or peg projecting at each end to rest in a groove so that it spins freely and deters the birds from perching. A simpler alternative, sometimes used, is to stretch a plastic coated curtain wire along the trough.

One of the disadvantages of troughing is that it needs to be replenished manually each day and any feed left overnight is an open invitation to rats and mice. Suspended tube feeders overcome this problem. Made in metal or plastic, they consist of a tubular central container into which mash or pellets can be put, with a surrounding trough below. A single tubular feeder of 38 cm (15 in) diameter will provide sufficient feeding space for twenty five layers. And holding about 14.95 kg (33 lb) of feed, it will only require topping up every five days or so for that number of birds. Adjust this type of suspended feeder so that the trough is at the level of the birds' backs.

Poultry-keeping in the back garden is not restricted to the concept of a house and run, of course. Earlier in this chapter the way to keep small numbers of birds in cages successfully was described, and this type of management is also adaptable to outside conditions, where the cages are roofed over and filled in at the back. But single-tier batteries take up a good deal of linear space, and many people opt for the neater compromise of a verandah, which gives the birds run space on a raised platform of welded mesh or wooden slats and keeps them out of the mud.

In its simplest form the veranda may be little more than a wide communal cage with a pent roof of asbestos, corrugated iron or plastic, having on one side to screen off cold winds. Wind has a discomforting effect on the birds in that it disarranges the feathers so they lose their insulating value. The whole thing may be built very quickly from 8 cm × 2.5 cm (3 in × 1 in) welded mesh on a framework of slotted angle iron with enclosed nest boxes at one end and feeders and drinkers at the front. Under some circumstances this may be sufficient, but many people favour an enclosed roosting area, feeling that the rather exposed situation of the birds, especially in cold conditions, or where foxes and other predators are a nuisance, may cost quite a number of eggs in a season.

The veranda, in fact, can usefully replace the run on a conventional house. Standing 60 cm-1 m (2 ft-3 ft) above ground level (the higher it is the more space there is for 'mucking out' manure at the end of lay) it allows the birds to be easily seen and serviced. Feed troughs along the front are often pivoted, tipping outwards for filling and inwards for the birds to feed.

It is just one step from this to the enclosed deep litter house and intensivism. Such a system, however, employing all the techniques described above for the 1.8 m × 1.2 m (6 ft × 4 ft) house on a larger scale, is really the first step to commercial production.

Looking beyond the conventional layer, several other forms of poultry can be fitted into the back garden system. Bantams, for instance, are well catered for in the types of housing described, and although they are poorer layers, nearly twice as many can be housed.

Ducks come a close second to fowls in popularity in small-scale poultry units. They are easily managed without making great demands on space. Housing is mainly a question of night protection for both birds and eggs against predators. A good, dry bed of litter and a non-stuffy atmosphere are the principal requirements and hen-house designs do not need much adaptation to be suitable for ducks, although space allowances of 0.28 sq m–0.37 sq m (3–4 sq ft) per bird are normally recommended. Young ducks can be satisfactorily reared in verandas. A house of 2.4 m × 1.8 m (8 ft × 6 ft) with a pent roof giving a maximum height of 1.2 m (4 ft) at the front and 1 m (3 ft) at the rear would hold a breeding flock of ten Aylesbury ducks and three drakes (or the same number of ducks for egg production). The upper 30 cm (12 in) of the front should be open, filled with wire netting. This positioning supplies ventilation while protecting the ducks, who can be excitable, from outside disturbance. The floor can be of wood, or concrete, overlaid with a litter of straw or dried leaves and bracken. Alternatively, 2.5 cm × 2.5 cm (1 in × 1 in) welded mesh, raised perhaps 30 cm (1 ft) over concrete permits easy cleaning down. Nest boxes should be kept invitingly clean and dry, but floor eggs are not uncommon with ducks who mainly lay at night and these soiled eggs should be kept away from clean ones. During the day, ducks are mostly allowed to range freely, but if this is inconvenient in a small garden, they can be confined by a 1 m (3 ft) wire fence if they are wing-clipped. Water should be liberally provided, although no pond is necessary except for breeders. Feeders and drinkers can be as for chickens.

The quail as an individual bird is certainly small enough to suit back garden conditions, but it is hardly a bird for the casual hobbyist. Several varieties are grown, and while the Japanese quail (*Coturnix Japonica*) is probably the most widespread, the Bobwhite (*Colinus Virginianus*) is grown particularly in America for its meat. The growth-speed of the quail demands constant attention to production and marketing details and there is little point in growing it without also becoming involved in breeding replacements. Given the necessary time and enthusiasm, however, it is an intriguing challenge.

Japanese quail mature in six weeks and under good management will lay up to 300 eggs in 360 days. When killed for meat after a six-week growing period, they weigh around 142 g–170 g (5–6 oz) oven-ready.

Management of these small birds is easiest in cages. A fine wire mesh of the sort used on old meat safes may be needed for the chicks, but adult breeders can be housed in safes similar to tier brooders for chickens. Such cages may be set up in a suitable garden shed. Up to 170 birds per sq m (16 per sq ft) may be caged from hatching to six weeks, but colony cages will be stocked at 40 birds per sq m (4 per sq ft). In a typical cage management situation the chicks may be started in solid sided brooder cages with a fine mesh floor, under an infra-red lamp. Glass water founts can be used, but stones or glass marbles should be put in the tray to prevent the very lively youngsters from climbing in and drowning. After five days they will be transferred to growing cages, in batches of say twenty-five to a cage. Hessian mats or plastic mesh will be placed over the wire floor to give them a foothold and again the sides are filled in, with hardboard for instance, to protect the birds from draughts and prevent them from escaping. At three weeks of age they are large enough to be allowed access to feed troughs fitted to the outside of the cage and here they remain up to slaughter at six weeks.

For anybody with access to a building offering the right space and conditions, however, quail can be kept intensively, like miniature game broilers, reared on sawdust litter, although cannibalism is a potential problem when birds are confined. A pen of $3 \, m \times 6 \, m$ ($10 \, ft \times 20 \, ft$) will hold up to 900 quail—just over 50 per sq m (4 per sq ft). For the first ten days they are reared under brooder lamps in a hover surround, then given the run of the pen. Because of their small food intake a proprietary ration of 26% protein is used by successful growers, though it is also possible to use turkey starter and grower type feeds.

As a less usual choice for the adventurous, the guinea-fowl has quite a lot to commend it. Well-known for its meat qualities, this bird also lays eggs averaging 42 gm ($1\frac{1}{2}$ oz) in weight which are highly praised for their culinary value. And in its own right, the guinea-fowl is a fine ornamental bird, with, in most strains, beautiful, delicately spotted plumage.

I said 'for the adventurous' but that is perhaps misleading, because this is a hardy species which responds well to good management and apart from the need for ample ventilation and a high protein diet—they are prodigious insect eaters on range—presents few complications. Small numbers can be kept under much the same conditions as chickens, in a house with a run or in verandas. Indeed, some poultrymen run guinea-fowl with chickens, though that sort of compromise generally means getting less than top performance from either type of bird.

Guinea-fowl are more normally kept on range, and if they are going to be confined, they should be reared up to it, rather than bought in as adults. Allow 18 cm–20 cm (7 in–8 in) per bird of roosting space in a conventional poultry house, and around 1.85 sq m (2 sq ft) of floor space for each. The house should be wire-fronted, because these are predominantly open-air birds. If they are wing-clipped, a 1.5 cm (5 ft) fence will contain them outside, where stocking rates of 500–700 to the acre are common.

Under natural daylight the laying season extends from April to October in the northern hemisphere but year-round production is possible with artificial lighting. Conventional nest boxes, at floor level and secluded, are suitable for housed birds. On range they make their own, well-concealed nests.

Young guinea-fowl, known as keets, can be successfully brooded with a hen foster-mother in a coop, or raised under a floor brooder, using the same techniques as for chickens (see Chapter 6) except that their small size must be taken into account and no mesh larger than 12 mm ($\frac{1}{2}$ in) used for surrounds and cages.

The breeder who decides to propagate ornamental guinea-fowl has a wide choice: Pearl and Lavender varieties, Whites (which offer the advantage of white meat, against the normally dark meat of the other strains), White Breasteds, Royal Purples, and Dundottes, which are a combination of Pearl Lavender and Royal Purple strains, are all available to him.

Monogamous in the wild, the male will mate with more hens in domesticated surroundings, a ratio of 6:1 providing good fertility. There is not much in appearance to distinguish the sexes, other than slightly bolder and heavier features in the male, even at maturity, but a clear give-away is their call, a repeated two-syllable sound from the hen and a single cry from the male. It is the noise of the guinea-fowl that can make it a serious trial of neighbourly friendship, however. If at all alarmed, both sexes can emit a piercing shriek, which has gained them a reputation as 'watch-dogs'.

For meat purposes, guinea-fowl reach eating weights at eleven to sixteen weeks, depending on the breed and whether they are grown intensively on litter, like broilers, or allowed to range. Modern selected strains are capable of reaching 1.8 kg (4 lb) liveweight, although a more usual selling average is 1.4–1.6 kg (3–3$\frac{1}{2}$ lb). There are two schools of thought about range-rearing, its supporters pointing out that, although growth is slower, the actual cost of feed is less because of the guinea-fowl's consumption of worms, grubs and insects. The other argument is that its greater age lends it more flavour. Nevertheless, the commercial future for guinea-fowl seems destined to follow a 'broiler' pattern.

Despite their relatively large finished weight, turkeys do not need elaborate housing, and a few can be kept in a small area without much difficulty. In a temperate climate they will live quite happily in outdoor pens with no form of shelter, but this calls for land, so that they can be moved when the grass gets wet or worn. Turkeys must have clean ground. It also involves wing-clipping and good fences to keep predators—and possibly poachers, at bay. A basic system of housing and a deep-litter floor is the answer where land is limited. This is the pole-barn, a simple structure usually with a pent roof of boards and felt, corrugated iron or asbestos, supported on a strong wooden framework, but with open walls of wire netting. Many variations are worked on the theme, and it can be as large or as small as required.

By covering the open wire with polythene sheeting in the early stages, it is possible even to rear birds from day-old in a pole-barn, although

Turkeys do well in open-sided 'pole barns'

they are normally transferred to the house after the initial brooding stages have been completed elsewhere.

With turkey production, an early decision must be taken on one's purpose in growing them. Is it to provide a large bird or two for Christmas or some other festive occasion, or is it to produce useful, medium-sized stock for the freezer, that can be eaten at any time during the year, with perhaps some surplus for sale to cover the overheads? On this decision will rest the size of housing you set up, because turkeys have been bred, in recent years, to fit market requirements, some fattening early, to provide well-finished lightweight carcasses, while others mature more slowly and economically to larger weights. Small turkeys reach 2.25 kg–3.2 kg (5–7 lb) oven-ready at about eleven weeks under commercial conditions. Medium strains go on to about 5.45 kg–6.35 kg (12–14 lb) at sixteen weeks while breeders and catering strains are getting up to 13.6 kg (30 lb) and more at twenty-five weeks. Clearly much more room will be needed for any programme that takes turkeys beyond sixteen weeks of age, although in general hen turkeys make lighter weights than stags—and incidentally create less noise, which probably makes them a good choice for the domestic grower.

Broadly, the space requirements for turkeys at different ages are:

age in weeks	square metres	square feet
0–4	0.09	1.0
5–8	0.14	1.5
9–16	0.37	4.0
16+	0.46	5.0

Assuming your intention is to grow medium-weight turkeys, therefore, it would be possible to keep six in a 'barn' of 1.8 m × 1.2 m (6 ft × 4 ft) up to sixteen weeks.

The site selected must be as dry and well drained as possible. To assist the process, dig it out to a depth of 5 cm–8 cm (2–3 in) and introduce rubble and clinker, tamped down to a level surface. The main house frontage should be orientated to make the best use of available sunlight.

With the framework established, allow the roof a good overlap to provide shade and prevent rain from soaking the litter. The lower sections of the walls can be built up to a height of 60 cm (2 ft) with weather-boarding, or any readily available panelling material. A lightweight door will also be needed.

Chopped wheat-straw or woodshavings make good litter for turkeys but if shredded paper or something else is easier to come by locally, use it, providing only that it gives the proper absorbency and does not create dust, to which these birds are particularly sensitive. Spread the litter to a depth of 8 cm (3 in) initially, and top it up with a further covering whenever it shows signs of becoming wet or compacted. It must be totally cleared out and replaced between batches.

Turkeys like a perch, although they will not die for the want of one, and a smoothed wooden bar 6.3 cm ($2\frac{1}{2}$ in) square with bevelled upper edges, will suit them very nicely. It should be about 60 cm (2 ft) above the floor and if two or more are required they should be at a similar height, to avoid 'top bunk' arguments.

A single tube feeder will be adequate for up to a dozen birds, or a trough may be used. A clean water supply is essential.

Once again, for those wishing to take turkey production a stage beyond the basic, the veranda can be a useful piece of equipment. One of the best known systems of this type was designed by the late William Motley, a leading breeder and rearer of turkeys from the 1930s to 1960s.

A typical veranda for, say, thirty turkeys, will measure 5.5 m × 1.5 m (18 ft × 5 ft) with a floor height of 61 cm (2 ft) and be made of timber and wire netting. The floor may be wooden slats or welded mesh and the roof of timber and felt, or possibly just felt laid over a wired framework. In an apex design, it will extend in height above the floor from 76.2 cm ($2\frac{1}{2}$ ft) at the eaves to 1.68 m ($5\frac{1}{2}$ ft) at the ridge.

There is an element of risk in using wire or slatted floors for turkeys, however, in that these heavier bodied birds are prone to foot troubles and breast blisters. In the case of breast blisters, even hard-packed litter might sometimes be a contributory factor occasionally, although their precise cause is still debated. The only remedy is to check both the birds and their conditions regularly.

Farm systems

Nothing illustrates more clearly the changes that have come about in poultry production since the 1950s than a comparison between the methods in use then and the specialist industry that has grown up today. At that time, commercial poultry-keeping was still within the scope of the general farmer. It needed a bit of land for the free range and fold systems of the day and the virtues of poultry manure on pasture were often extolled as a reason for keeping birds in this way. Hen yards and deep-litter houses were steps towards intensivism, and some large battery units were already in existence, but a gulf separates even those from their modern counterparts.

Certainly for egg and broiler production on the large scale, the old systems are totally uncompetitive. There is no way in which they can provide the centralized services and labour efficiency of the well-run intensive system to achieve optimum performance throughout the year. Sad, unromantic, but true, and you owe the low cost of your breakfast egg to the development.

Free range means just that; birds housed in scattered sheds, with no wire netting to restrain them. Field houses, on skids or wheels to enable them to be towed by tractor, provide nesting and roosting accommodation for about forty layers each. Measuring $1.8 \, m \times 2.4 \, m$ (6 ft × 8 ft) and upwards, such houses are still obtainable.

Slatted floor arks can also be found in makers' catalogues and perform a similar function to the solid floored houses, except that perches, dropping boards or a droppings chute are dispensed with and more birds can be housed in a given area.

As already noted in this chapter, free range is still practicable for guinea-fowl and geese, and even ducks and turkeys may be raised in this way to meet a specific local demand. At favourable times of the year, access to range lessens the feed bill and the manure enhances the sward. But for commercial egg production or broilers it is an economic non-starter against up-to-date systems. It involves a lot of work in feeding and attending the birds and a lot of land, for a thin covering of birds—sometimes no more than fifty to the acre—that can be put to better use.

Marginally better is the fold system, in which flocks of about twenty-five adults or fifty growers are contained in small slatted-floor houses with a run attached and moved systematically to fresh grass each day. This supplies the benefits of manure distribution on the fields while giving a higher concentration of birds per acre. Breeding families can be kept in individual folds. But again, although folds present a cheap method of housing, labour costs and land usage are against them. When widely used they were criticized for the high proportion of dirty eggs they produced and the difficulties of movement on heavy ground. Freeze-ups were common in winter and feather-picking could be a problem with both layers and fattening turkeys, and cases of cannibalism and damaged backs were also occasionally reported.

Semi-intensive systems centred on fixed houses with an enclosed run but usually a fairly generous floor allowance of 0.28–0.37 sq m (3–4 sq ft) per bird, so that birds could be totally housed if necessary. A typical semi-intensive house for 100 layers would be about 8 m long and 4 m wide (26 ft × 12 ft), generally with a span roof having a protected opening at the ridge, to obtain optimum airflow, since no fans or forced air systems were used.

The hen yard principle is a shed with a strawed run in front of it, popular with general farmers when it was commercially economic, because many farm buildings could be adapted to hold hens. Several hundred pullets could be kept in this way. Within the shed, a droppings pit is constructed with a dwarf wall 60 cm (2 ft) high, its size determined by the length and number of perches above it, made up as a removable frame. Wire mesh is stapled to the underside of the perches, to prevent the birds from getting down into the pit. Communal nest boxes, feed troughs—suspended to make them rat-proof—and water troughs, as described earlier in this chapter, complete the equipment.

Commercial management

The most obvious difference between commercial poultry housing and that of non-commercial producers is its size. Intensivism has brought specialized buildings capable of maintaining tens of thousands of birds in individual productive units.

In the variable British climate, intensive houses have gone all the way in environment control. They are built on substantial foundations, generally with concrete bases. Almost by tradition they are timberclad with asbestos or composition roofing, although aluminium cladding has recently put in an attractive appearance and is favoured by manufacturers who have also found this a successful material in hot climates, where its reflective qualities are helpful in keeping the interior cool.

Elaborate ventilation systems have been developed to cope with the considerable demands of thousands of birds in an enclosed space and these are described in Chapter 8, but good control depends on what lies just under the skin of the buildings.

Insulation

The essence of good insulation is that it should help to maintain an ideal temperature in the house regardless of external conditions. To do this, it must provide an effective barrier to thermal changes. Materials vary widely in the speed at which they transmit heat. The aluminium cladding mentioned above, for instance, although it reflects heat, is a fast conductor of temperature changes and would be useless on its own. It must be supplemented by an insulating lining. Air, on the other hand, is a good insulant. A well sealed cavity is often a feature in roofs and walls, but more usually the main barrier will be a foamed or cellular material incorporating a lot of air in its structure.

Two measurements are used to gauge insulation efficiency. For complete thickness of walls, roofs or floors taking into account all their various claddings, air spaces and liners, a 'U' value is applied. This gives the number of British thermal units passing through 1 sq ft of the structure when the difference in air temperature between one side and the other is 1° F (it abbreviates to Btu/ft²/hr/° F) one Btu being the amount of heat required to raise the temperature of 1 lb of water through 1° F. The lower the number—i.e., the fewer Btu's penetrating the structure—the better the insulation. Related to this is the second measurement, the 'K' value, which tells the efficiency of individual materials. Again it records the Btu's passing through 1 sq ft in 1 hour but this time it is based on a standard thickness (1 in) of the insulant. So 'K' value is Btu/in/ft²/° F.

In a metric age these values have to be convertible of course and Imperial 'U' × 5.678 gives the metric equivalent, while in the other direction, multiplying metric 'U' by 0.176 gives Imperial 'U'. Metric 'K' is found by 'K' × 0.1442 and Imperial 'K' by metric 'K' × 6.934. Some representative 'K' values are shown in the table.

Table of Imperial 'K' Values

Material	'K' Values
Aluminium sheet	1,420.0
Asbestos cement	2.5
Asbestos partition board	1.75
Brick	8.0
Chipboard	0.87
Concrete (dense)	5.0
Concrete (aerated)	1.0
Fibreboard	0.35
Glass fibre	0.25
Mineral wool	0.30
Roofing felt	1.4
Rigid polystyrene	0.24
Rigid polyurethane	0.16
Rigid urethane	0.16
Straw (compressed)	0.59
Vermiculite (loose granules)	0.45
Wood	1.00
Wood wool slabs	0.55

Looking at these measurements in practical terms, a modern broiler house in British climatic conditions will be expected to have a 'U' value of around 0.08 and that of a laying unit about 0.1. These are quite high efficiency levels, even compared with similar houses in the early 1970s, but they are repaid by the savings in fuel (for brooders) and feed, items which have soared in cost in recent years. It is a false economy to skimp on insulation in commercial units.

House structures are built up in layers for several reasons. On the

Commercial deep pit house. Caged layers live in upper, wooden structure. Lower wall encloses pit

roof, the largest and therefore most important area of heat loss or gain, the external cladding must be hard-wearing, weather-proof, requiring little maintenance and reasonably light in weight. Where high daytime temperatures can be expected, its heat reflective properties may also be important. Ribbed aluminium, corrugated asbestos sheeting or corrugated bitumenized sheeting (Onduline) are common choices. The latter can be painted in light colours, if required.

Within the cladding there may be an air space, then a layer of insulation—usually polystyrene or glass fibre—then, very important, a vapour barrier to prevent humidity in the house from saturating the porous insulant and ruining its insulating qualities. The vapour barrier is normally polythene film, and glass fibre can be obtained already in envelopes of polythene. Finally, to support everything in place, a rigid lining of asbestos cement or possibly oil-tempered hardboard completes the sandwich. Such a roof, consisting of outer cladding, air space, 5 cm (2 in) polystyrene, vapour barrier and inner lining, will have a 'U' value of 0.08. Alternatives may use different materials, different thicknesses of insulation and dispense with the air space, but the same basic sandwich system will apply. One popular variation, however, is the 'attic' layout, in which a ceiling about 50 mm (2 in) thick is introduced above the birds, and the roof itself is simply lined with polythene or foil. In this case fans in the attic drive air down through the glass fibre ceiling, which is usually supported on a framework of wood and wire mesh.

Walls may be built solidly in traditional materials, particularly where a farmer has a possible secondary purpose in mind for the house should his poultry business fail, but it is worth noting that a 23 cm (9 in) brick wall with 19 mm ($\frac{3}{4}$ in) rendering only has a 'U' value of 0.41 while 15 cm

(6 in) concrete insulation blocks rendered inside and out give a 'U' value of 0.20. Again a lightweight sandwich construction is more common.

A lightweight sandwich construction is likely to be both cheaper and more efficient. The timber-clad wall, for instance, having 2.5 cm (1 in) nominal tongued and grooved boarding, 2.5 cm (1 in) of glass fibre, vapour-sealed on both sides and an inner skin of 9 mm ($\frac{3}{8}$ in) plywood presents a 'U' value of 0.14. Higher efficiencies can be achieved with thicker insulation, whether with wood, metal or asbestos cladding.

Do not despair if you have an older house relying on a simple cavity for insulation—but equally, do not be complacent about it. The same sort of service that has been available for domestic houses for some time can now be employed to inject urea formaldehyde foam into the cavity and bring the house up to modern standards. In fact, simple shell buildings can also be treated with rigid urethane foam which is sprayed directly on to the inner surfaces.

Insulation factors are only part of the story when considering materials. Cost, durability, ease of installing, fire risk, resistance to pressure washing, pecking and insect attack—beetles sometimes ravage polystyrene—should all number in the pros and cons.

The key to modern commercial management is really summed up in one word—convenience. Deep litter has persisted in use for broilers, turkeys and breeding birds in the main since the early days of large poultry houses but batteries were quickly recognized as the most convenient way to house laying stock. The birds could be seen readily. Eggs were collected from the trays in front of the cages without difficulty, and non-layers could be quickly identified. Floor-space could be used to its fullest advantage, and the general ease of management more than outweighed the high initial cost of the cages. All these old arguments still hold good and a few more have been added since.

Starting with 'cafeteria' batteries, in which food and water troughs on a gantry were propelled slowly along the row of cages, the pursuit of labour-saving devices and layouts has continued relentlessly. There is a Japanese system where cages round a central column rotate slowly to bring the birds to the feeding point.

Automatic egg collection was an obvious candidate for development, but while the principle of a moving belt along the front of the cages was simple, bringing all the eggs to a common level from the different tiers without breakage proved a stumbling block for cage designers. Partly as a result, in the 1960s and early 1970s, large flat-deck units enjoyed a vogue. With this system all the cages are on one level, so that bringing the eggs from the birds to a cross-conveyor carrying them direct to an adjacent packing station is no difficulty. It also has the advantage over vertical tier units of providing a uniform lighting pattern for the birds, illuminated from above, and a clear drop for manure to a pit below. Its disadvantages, however, are a heavy reliance on the working of automatic feeding and watering devices, the difficulty of seeing and reaching birds, access to which is from above the cages, via catwalks, and a lower stocking density than is possible with multi-tier systems. And when the engineers finally cracked the problem of egg 'escalators' that did not crack eggs, emphasis swung back to tier batteries.

One chore that took on new dimensions with the advent of cages was manure handling, which has had its own influence on both cage layout and house design. Cages built in straight-sided vertical tiers suffer the obvious disadvantages that droppings from the upper levels must be prevented from falling on birds underneath and must be removed from between the tiers either manually or mechanically. Several methods of doing this have been developed and are discussed in Chapter 9, but it does involve a regular handling process which many poultry-keepers would rather do without. Vertical cages can also be difficult to light evenly. On the other hand, the upright stack provides the most compact layout and the greatest stocking potential. As building and feed costs rise, therefore, this layout, which offers the lowest housing cost per bird and easy maintenance of warmth is becoming, if anything, more popular.

Three-tier systems are the most common, because these enable birds in the top cages to be seen and reached relatively easily, but some four- and five-tier layouts do exist, and allowing for obvious questions of weight—1,000 adult layers represent a total load of about two tons—the only real objection to going higher is the inaccessibility of birds. This might be overcome by some form of travelling gantry for the stock-man.

Vertical cages are readily adaptable for rearing, just by the addition of fine mesh plastic floor mats and special fronts to prevent the escape

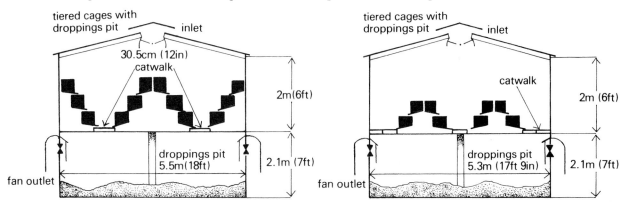

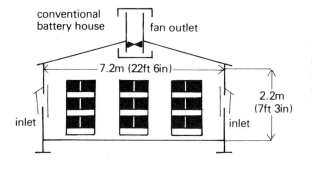

Typical house and cage systems in use today include three and two tier stepped cages over deep pits (*top*) and pre-fabricated, conventional houses with vertical cages (*left*).

of the chicks, and manufacturers can now supply cages in which birds can be kept from day old to the end of lay.

Broilers can also be raised in cages, which owe a lot in their design to the old tier batteries used in earlier times for rearing. A number of birds, perhaps thirty or more, are run together in communal cages. But a high incidence of breast blisters set the designers thinking, until they introduced 'trampoline' floors of springy plastic, perforated with holes of about 2.5 cm (1 in) in diameter, which helped to cure the problem. High costs of capitalization, and the strongly entrenched methods of floor production of broilers have so far prevented broiler cages from becoming widely used however, despite the superior use of house space they offer. Their running advantages are elimination of litter costs and coccidiosis problems, marginally better weight gain because the birds do not move about so much and easier catching. Some manufacturers have even developed cages which form transportable modules, so that the birds can be taken straight to the slaughter line without being transferred to special travelling crates. A highly integrated production system, in which the grower and processing plant are closely linked, is essential for this to be workable, however, and again the practice is not widespread.

Through efforts to cut out continuous manure handling, the Californian layout of stepped cages evolved. Tiers are offset so that manure can fall straight to the floor of the droppings pit below, where it may be left for five years if necessary and the arrangement works well. It has the additional advantage that birds on all levels can receive uniform lighting without difficulty. The loss comes on its use of house space. The spread of tiers means that fewer cages can be accommodated for a given house area, against a four-tier vertical system, although a saving can be made on the width of catwalks. Semi-stepped cages provide a compromise. Here the tiers partially overlap, the occupants of the lower cages being protected by a sloping plastic sheet which wards off droppings from above.

Provision of a droppings pit is another drawback when it comes to initial house costs. Sufficient headroom must be allowed in the pit for a tractor to enter for cleaning-out duties, so the support walls need to be about 2.1 m (7 ft) high. An excavated pit, with the house at ground level, would not only be expensive to construct and keep dry, but difficult to clean out. And ventilation outlets could not be arranged in the sides as is normal with the pressurized down-draught air flow used with droppings pit systems. However, it is possible to convert existing, solid based houses to this principle, by jacking them up hydraulically.

Under commercial conditions, water is almost universally piped to cages now by plastic tubing, into which are screwed metal nipples. The line may run externally along the front of the cages, or be placed through the centre of them. Generally separate header tanks, with a ball-valve, serve each tier. They are placed at the end of the cage-row and the tubes run from them along the house.

Nipple drinkers have a small projecting pin, easily pushed in by the bird's beak to open a valve and release water. One or two drinkers for every four birds must be the rule. Some poultrymen prefer cup drinkers. These use the same type of supply line as the nipples, but are small open plastic cups, fitted with a trigger which the bird simply pecks to allow water into the bowl.

Having an enclosed watering system brings several benefits. It eliminates the evaporation experienced with the old-style open troughs, and this can be important in winter, when moisture evaporating in a house creates a heat demand and brings down house temperature. It reduces wastage and needless splashing; and open trough flowing waterers often presented a problem in disposing of a large amount of overflow water, leading to corrosion of metal cages and feed troughs. It prevents feed and contaminants from getting into the water supply. It deters the formation of algae. On the debit side, if there is a blockage in a water line it is not so immediately obvious as a dry trough would be. Faulty drinkers in individual cages can go unnoticed and either leak or—more seriously, fail to deliver water. But, of course, it is the test of a good manager to watch and remedy every malfunction, and inspecting water lines is one of the most vital jobs. Header tanks must be at the right height, as recommended by the drinker manufacturer, for drinkers to function properly; 5 cm (2 in) too high and some drinkers will leak. Every month or so systems should be flushed through, under pressure, to clear out residues.

Wherever large quantities of feed are being handled these days, suppliers prefer to deliver in bulk. This eases their work-load. Prepared rations are simply transferred from the mill to the bulk-tanker vehicle which transports them to the farm. Here it is a simple matter for the driver to run a pipe from the specially equipped lorry to the permanent bulk bins standing outside each poultry house, and auger or blow the feed into the bin.

Bulk bins are made in metal, wood, glass fibre, 'space-age' plastics and even soft fabric, and all have their good and bad aspects. The main enemies of feed stored in bulk are damp resulting from leaking containers or condensation in the bin and temperature extremes.

Bridging—clogging of the feed across a narrow part of the container so that it does not flow—can also be a problem. Some producers make a habit of bouncing a ball against the bins to shake the feed down.

Important considerations in selecting a bin are that it should have adequate capacity for the flock it is supplying; good weather protection; easy visibility of the contents; good 'flow' characteristics and security against vermin. Then it is up to the stockman to ensure that it is regularly topped up and that no stale feed is left adhering to the sides. Again, good management here will stop at source many of the problems that bad managers always seem to be complaining of.

Feed is usually taken from the bin into the house by a cross auger which deposits it into hoppers. From these it may follow a number of different routes according to the management system involved.

In cage houses, continuous troughing along the front of the cages may be replenished by a travelling hopper, which moves along the cage row, depositing feed into the troughs on all tiers simultaneously. If the hopper is pushed manually it is a good opportunity for the stockman to inspect the birds as he goes, but commonly now the hopper is electrically operated and can be set to travel automatically several times a day by regulation of a time clock on a control panel.

Other methods of supplying feed to troughs include the use of an enclosed auger, and drag-chains. These can also be regulated by a time clock.

Nearly all feed troughs on cages are metal and they have two bad habits. They become deformed and battered, so that feed depth can vary from cage to cage, and they develop holes. Such defects are crucial. Not only are they an unkindness to the birds, who will be quick to show a reaction in performance, but they are directly wasteful.

Unevenness in troughing is compensated by special self-levelling devices in modern travelling hoppers, which will spread the same ribbon of dry mash along the trough regardless, but the only answer to holes is to block them. This can be done by patching with glass fibre, or inserting a complete plastic liner the length of the trough. The life of troughs can be prolonged, however, by stopping water from faulty or badly placed drinkers, saturating the feed. And the corrosion problem can also be overcome by using plastic troughing from the outset.

A well-designed cage trough has a lip on the outer edge and good depth to prevent feed being flicked out. Some systems have a chain or spiral in the bottom of the trough, also to prevent wastage. It is a useful maxim to feed little and often, so that there is never a great depth of feed to be churned over.

With the steady increase in feed prices a number of precision devices have come on to the market to ensure that birds receive just the correct ration allocation and no more.

In some systems, feed can be distributed manually, using feed hoppers suspended from overhead rails and pushed into the house. Feed is simply doled into troughs or tube feeders of the type described earlier. But it is far more common to employ either a continuous chain feeder, or a system of pan feeders, to which feed is delivered automatically through tubes.

Egg collection is another management task largely replaced by automation, although some producers deliberately collect by hand and combine this with their daily inspection of the flock. A simple, low trolley can be run down the aisles in a battery house and the eggs collected on to Keyes trays. With care, cracks can be kept to a minimum by this method. Alternatively, plastic-coated wire baskets may be used and are more convenient in a deep litter system.

The job is time-consuming however and a good deal can be saved with automatic belt collection. Breakages, with the latest systems, are

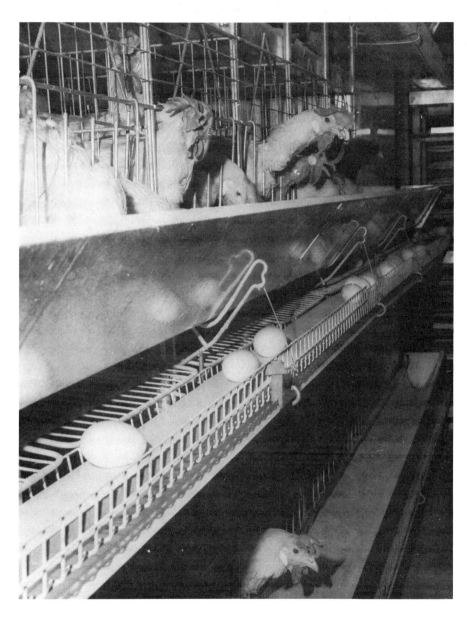

Most commercial eggs are now collected automatically on a travelling belt

not necessarily higher than with hand collection either. Maintenance is more likely to be a problem. Motors can go wrong, belts can stretch and occasionally the producer may find he is back to time-honoured methods while his expensive equipment stands on the side-lines.

Communal-type nest boxes in deep litter houses hardly lend themselves to mechanization, but even here one ingenious system, the Autonest has been developed. A long tunnel nest, with entry holes at intervals along one side provides a comfortable bed of buckwheat husks for the hens, who lay quite happily in it; but the nest material is on a conveyor. When collection time comes round the belt is set in motion, the material circulates to a collection point, where the eggs are gently 'combed' out and the nesting material returns to the nest area.

8 The controlled environment

What is environment control? In a sense, the moment you confine birds under a roof you are controlling their environment. Rain and snow no longer fall on their heads. Light comes to them from different angles. Air patterns are changed. Taking it one step further, however, it can be argued that bringing chickens to a cool climate in the first place was a retrograde form of 'control', since under natural conditions they are warm-weather birds.

Perhaps some thought about the original environment of the bird would have short-circuited a little of the painstaking research that has gone into keeping birds in enclosed houses because they are found to do rather better at temperatures around 21°C (70°F) than they used to outside in a chilly climate. Early users of intensive systems learned, little by little, the value of insulation, the principles of ventilation, the adaptation of lighting and correct nutrition. Scientists and technologists helped to determine the proper environmental requirements and along the way old problems such as the winter drop in egg production previously accepted almost as a matter of course, were overcome.

Among the lessons which had to be learned was the crucial part which air changes play in any building housing a large concentration of birds. The stuffiness of the overcrowded room is familiar to everybody. It results from the body-heat and moisture given out by all warm-blooded creatures, and in the case of birds it is accentuated by their higher blood heat of around 41°C (106°F). In a well-insulated building containing say 5,000 adult birds, the build-up of heat and humidity is rapid, if not tempered by regular air-changes. Indeed, failure of ventilation in a house containing many birds is likely to lead to heat prostration and deaths within twenty minutes or so, if no inlets are provided.

Too much air lowers the temperature, leads to draughts and increases food consumption; too little will step up temperature—potentially to disaster levels, as stated—and generate irritant ammonia gas from droppings. Air changes are clearly important therefore, but how often should they take place?

When environment control was in its infancy, ventilation rates were fairly arbitrary, six to eight changes an hour being normal. Inlets for natural ventilation systems were a recommended 3.5 sq ft per bird, and double the extraction area. As house size grew and fan ventilation was introduced, more complex formulae came into being. Initially a maximum ventilation allowance of 1.5–2 sq ft per minute for every pound liveweight of birds in the house, and a minimum of 0.5 cfm/lb were laid down. The standards were revised when it was realized that the acceptable maximum of 6–8 cfm which this gave a 4 lb layer translated into 45–60 cfm when the formula was applied to a 30 lb turkey. It was then changed to one based on daily food intake. To obtain an indication of maximum ventilation rate, this multiplies each pound of food eaten in a day by 25 cfm. A bird on 4 oz/day therefore receives a quarter of 25 cfm, which is still just over 6 cfm, but the heavy turkey, eating perhaps 8 oz/day now receives a more acceptable maximum of 12.5 cfm. At the other extreme, minimum rates were set at 2 cfm/lb feed/day. Fortunately the new standards relate closely to the metric m^3/second/tonne/day, 25 cfm/lb/day being equivalent to 25 mstd and 2 cfm/lb/day matching 2 mstd.

Some guide figures are shown in the table, but it should be remembered that they are only a guide, and no substitute for the nose and eyes of a good stockman. If ammonia levels are high or birds are panting, it is too hot, whatever the records say.

Ventilation rate requirements

STOCK	Weight		Maximum requirement		Minimum requirement	
	kg	lb	m^3/h bird	cfm/bird	m^3/h bird	cfm/bird
PULLETS AND HENS	1.2	2.5	10	6	0.8–1.3	0.5–0.75
	2	4.4	12	7	1.3	0.75
	2.5	5.5	14	8	1.5	0.9
	3	6.6	14	8	1.7	1
	3.5	8	15	9	2	1.2
BROILERS	0.05	0.1			0.1	0.06
	0.4	0.8			0.5	0.3
	0.9	2.0			0.8	0.45
	1.4	3.0			0.9	0.5
	1.8	4.0	10	6	1.3	0.75
	2.2	5.0	14	8	1.7	1
TURKEYS	0.5	1	6	3.5	0.7	0.4
	2	5	12	7	1.2	0.7
	5	11	15	9	1.5	0.9
	7	15	20	12	2	1.2
	11	25	27	16	2.7	1.6

Although it may seem odd at first to use feed intake as a measure of ventilation requirements, it is really a logical approach since the metabolizable energy in the food goes partly towards body heat. In fact there is a delicate balance between the ME taken in and the egg output, weight gain and heat loss of the bird.

Particular attention has been paid to this in experiments at the ADAS Experimental Husbandry Farm, Gleadthorpe, where it has been found that appetite falls by about 1.5% for every degree C (1% per °F) rise in temperature (or 4 k cals per °C) between 18–24°C (65–75°F). The explanation is that as environmental temperature goes up the bird needs to use less energy to maintain body heat, so it requires less food. Some caution is necessary here, however, because the smaller food intake could well lead to dietary deficiencies unless the ration is adjusted accordingly, and advocates of the system generally suggest raising the protein level from say 15% to 18% for layers.

Obviously young birds have different heat requirements, as described in Chapter 6, but on the basis of these calculations the optimum temperature for adult layers seems to be close to 21°C (70°F). Broilers also appear to benefit from temperatures in the 18–24°C (65–75°F) range, by improved feed conversion, though again nutrient density must be carefully watched, while turkeys—birds which favour a cooler climate and can still be grown in open pole-barns—have shown improvements of up to 6 k cals per bird/day/°C rise from 10–21°C (50–70°F) with an optimum in the region of 16–18°C (60–65°F). Less is known about the precise requirements of other species, but intensively kept guinea-fowl are generally housed at 21°C (70°F) at a relative humidity of 75% while quail seem to thrive best at 18–21°C (65–70°F). Ducks are kept under controlled environment conditions, and these systems are just starting to be used for breeder geese.

Under temperate conditions, an insulated house with a full flock of birds should attain an optimum atmospheric temperature just by employing the body heat of the birds with ventilation rates of 2–4 cfm/lb feed/day. Air circulation is kept at the lowest level at which oxygen supply can be maintained without leading to ammonia, dust and humidity problems. How well do the various systems match up to these requirements?

Natural ventilation

The difficulty in operating natural ventilation systems, where no powered fans are employed, is their lack of precise control. Air enters through light-baffled intakes in the walls, or, if it is a house raised above ground level, through inlets in the floor. As it is heated it rises, to escape at ridge outlets. This ensures a steady succession of air changes as long as conditions outside are favourable, that is, relatively windless and with no extremes of temperature. Unfortunately, when temperatures are low outside, the contrast with the warmth in the house causes the air changes to accelerate just when the producer wants to retain heat, while hot weather cuts down the differential between air entering and air already in the house and reduces ventilation when it is most needed. In addition, wind pressure can play havoc with the stability of such a system, and light control is, at best, a problem.

Nevertheless, natural ventilation has several advantages that have encouraged technologists to look again at ways of adapting it to modern conditions. Energy-saving is the most obvious benefit. Simple convection does all the work without the use of fans. This also frees it from the dangers of total breakdown through power failure which threaten other systems.

To gain better control over air-flow, relative to conditions in the house, ADAS specialist Peter Spencer has introduced the idea of temperature-sensitive rams, as used by horticulturalists to regulate greenhouse ventilation. In one commercial example a raised house with capacity for 7,000 layers allows air entry at floor level through boxes which are closed and opened automatically by the hydraulic rams. These contain a chemical which is activated at certain temperatures, in this case 18-21° C (65-70° F), to maintain steady conditions in the house. The shuttered open ridge can be closed manually when required. This is probably as near to the ideal of controlled ventilation as it is possible to get with a natural system.

Other producers have found that sizeable flocks can be kept in buildings employing cross-flow ventilation without fans, if the houses are relatively narrow and sited across prevailing winds. I have in mind one deep pit house containing 5,000 birds in flat-deck cages which is 69 m long and only 6 m wide (226 ft × 20 ft).

This hit or miss arrangement, in which doors have to be opened on hot days, is unusual under British conditions, but open-sided sheds for both layers and meat birds are common in hot climates such as are found in America's southern states and in Australia. When temperatures drop, these houses are curtained with plastic sheeting. The ventilation technique in a cross-flow house is to introduce the main airstream as near the roof as possible so that it generates rolling secondary currents which bring fresh air down to the stock. There should be no obstructions in the roof to deflect the primary airstream on to the birds. Powered ventilation can be an asset in these curtained houses at certain times of the year.

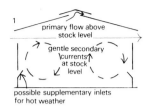

Natural ventilation
This and the following five diagrams show the effects of various ventilation systems

Powered ventilation

Introduction of fans to a house gives the poultry-keeper more independence from outside influences and greater flexibility of control. On the other hand it brings in a host of new factors, all of which cost money and can go wrong. Thermostats, thyristors, control boxes, timing devices, it is a long and complicated shopping list. Where do you begin? In fact most of the problems are taken out of the poultryman's hands by the specialists employed at the time a new house is erected or an existing one adapted, and the layman is well advised to leave electrical installations to electricians.

At least three fan types are manufactured, but by far the most common in poultry houses is the propeller. The work capacity of a propeller fan is influenced by its diameter, which is fixed when it is manufactured, by its speed of operation and by the pressure against which it is operating, both of which can vary. The table shows the capacity of two of the commonest fan sizes found in poultry houses, although other sizes can, of course, be obtained. From the figures it can be seen that a fan

Extractor fans built into wall of a Hallam deep pit house. Air is expelled through mesh-covered louvres

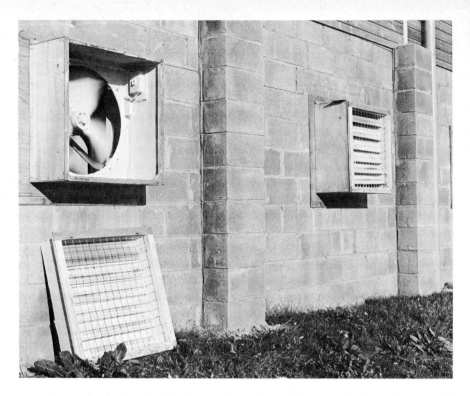

may become totally ineffective at less than full speed when faced with strong pressures, which can be caused by wind interference or narrow inlets. Air inlet design is therefore of special importance.

Rate of air displacement (cfm)

Fan diameter (in)	(mm)	Speed (rpm)	In free air	Light pressure	Strong pressure
18	457	700	2,000	1,740	—
18	457	900	2,600	2,350	—
18	457	1,400	4,000	3,850	2,600
24	610	560	3,700	3,000	—
24	610	700	4,600	4,200	—
24	610	9,000	6,200	5,800	3,400

Where total environment control, with windowless housing, is required, inlets must screen out extraneous daylight yet allow sufficient airflow, so they are boxed in and light baffles are introduced (see diagram). As air enters the house it has first to pass the baffles and these usually present a smaller aperture than the actual point of entry to the building. The smallest area is called the minimal free area and in orthodox systems should provide 0.05–0.08 m² per 1,000 m³/hr (1–1½ ft²/1,000 cfm) of maximum fan capacity. Delivery area—the point of entry to the building—is usually based on an allowance of 0.3 m²/1,000 m³/hr (5 ft²/

A wind-break fence
improves fan efficiency

1,000 cfm). There are exceptions to the rule, however, where special
high velocity systems are installed which employ very narrow inlet slots.

Wherever wind is likely to affect fan efficiency, and that is in most
cases with wall inlets or outlets, the importance of wind-breaks has been
increasingly recognized in recent years. Trees or bushes can be quite
effective, but a solid fence approximately 30 cm (1 ft) from each side
of the building and as high as the eaves is more certain. To prevent
funnelling of currents between the fence and the building, partitions
are built across at intervals.

Anti-backdraught shutters or flaps are another necessary precaution
against erratic ventilation in the house. Without them, fans which are
switched off may allow air through in the wrong direction, something
not wanted, especially in winter, when many producers adopt the policy
of sealing off some fans when no more than half the capacity of the system may be needed. It is better to run half the fans at full speed than
all the fans at half speed.

'Conventional' ventilation started with the installation of fans to
enhance the air flow in naturally ventilated buildings. In other words,
fans were placed at the ridge, to draw air upwards from the wall or floor
inlets, or they were positioned in one wall to draw air across from inlets
on the opposite side, in a cross-flow pattern. Such systems are cheaper
than more recent developments. They work in the medium velocity
range, broadly 0.25–2 m/sec (50–400 ft/min) and can still be found
world-wide. Unfortunately simplicity in design does not necessarily
mean simplicity in operation however. These systems offer no control

Roof extraction—wall
inlets: ridge outlets

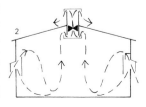

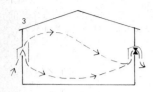

Cross ventilation—wall inlet: wall fan outlet

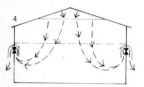

Attic ventilation—gable wall inlet: wall fan outlet

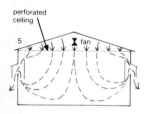

Pressurized ventilation through glass fibre ceiling

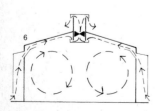

Narrow slot, high speed ventilation

over the temperature of incoming air, and winds can have a big effect on the speed of flow, which in turn will influence the pattern of air distribution in the house. It calls for a skilled stockman to attend to signs of increasing ammonia, damp litter or dustiness, and adjust the fans accordingly. Light control, where required, is also more difficult with systems of direct air supply.

Ridge inlet with wall extraction, putting natural airflow in reverse, has the advantage of making fans easier to reach for servicing, and saves on expensive roof fan housings. In deep pit units, the fans are set in the walls of the pit. But again there is no control over incoming air temperature or speed, and in summer the flow tends to follow the inner contours of the building and exit through the fans, while in winter cold air tends to plummet on to the birds below the roof inlets.

Pressurized ventilation was the first step towards overcoming problems of controlling air entering buildings. Fans were placed at the inlets and forced air into the house. This improved air distribution and light control, and led to the introduction of duct systems, a powerful bank of fans being mounted in one end of the house and blowing air—which could even be warmed or humidified if necessary—along perforated ducts of polythene or pegboard above the birds.

Ducted systems introduced a new approach to ventilation which is now very popular in England. It was found that air could be brought into the house and distributed to the birds at low velocities of less than 0.25 m/sec (50 ft/min) by fitting a porous ceiling of about 5 cm (2 in) of glass fibre in the house.

The fans are fitted in roof-mounted shafts in the loft or plenum chamber, and as pressure is built up, the air filters down—literally, for it is estimated that 85% of dust particles are removed—through the ceiling. Existing houses can usually be easily adapted to the system, which is commonly used for caged layers and rearers, although some broiler-houses now have it. Its biggest single drawback is that, if power-cuts occur, stifling temperatures can build up very rapidly. Special fail-safe devices are essential. They are generally trap-doors held by electro-magnets which fall open if electricity supply stops, allowing natural air convection to take over. Another difficulty is the need to replace the glass fibre periodically. Ceilings will last several years without clogging in reasonable climatic conditions.

While air from the glass fibre ceiling may come in at no more than 0.1 m/sec (20 ft/min), low velocity systems are not the last word in ventilation. It seems likely that that will not be written for some time, because at the other end of the scale, high velocity systems are available. These use fans at full speed to draw air into the house through narrow slots at velocities of up to 7.5 m/sec (1,500 ft/min). At these speeds, the airstream clings closely to ceiling and wall surfaces until it becomes mixed with the warm air already in the house and circulates to the birds.

Main benefits of the high speed systems appear to be the stable ventilation they introduce to the house, regardless of windy conditions outside, and a possible saving of fan-wear by using constant-speed rather than variable-speed controls.

From Denmark another option which has become popular particularly in cold-weather areas, combines air inlets and outlets in the same roof-mounted 'chimney'. Cool air is injected into the house through adjustable nozzles at 10 m/sec (2,000 ft/min) and stale air is removed under precise control which allows for recirculation of warmth in the house. Economic retention of heat is welcomed, particularly by broiler producers. One of these units is equivalent to several conventional fans, but they are correspondingly expensive to buy.

Controls

To keep house temperature stable at a selected level calls for control over ventilation rate, and this is achieved by switching fans on and off, or by varying their speed. The principle is that ventilation rate is kept as low as possible—with very young stock probably not used at all for the first few days—and fans are only brought into play when sensors in the house react to a rise in temperature. As warmth increases, more fans start up or, if on a variable setting, increase their speed.

Because their weight, and therefore their heat output, is relatively unchanging, adult birds are fairly easy to cater for. The minimum for 2 kg ($4\frac{1}{2}$ lb) hens will be set to 1.3 m³/hr/bird (0.75 cfm/bird) for example, and if house temperature increases it will be automatically countered by increased ventilation. With growing birds, weekly adjustments of controls will be needed as the minimum air requirement increases.

Controls, it should be noted, react only to temperature. It is still up to the poultry-keeper to ensure that ammonia and humidity levels are kept in line by regular visits to the house.

Of the control methods mainly used for livestock houses in the UK, on-off control is the simplest and cheapest. It involves a thermostat, which is a temperature-sensitive device activating a switch to bring in a fan whenever temperature reaches a certain level. Several fans in the house may be thermostatically adjusted to come on at different temperatures, one running to maintain, say, the minimum rate of 10%, with others switching on if the temperature rises. A technique now used is for one fan to come on automatically for a timed period of several seconds in every ten minutes to provide the necessary background air movement. If conditions in the house change, the timed period can be lengthened or shortened accordingly. The other fans are regulated in the usual way by thermostats.

Proportional controllers use a different system, changing the speed of the fans rather than switching them on and off. They are the commonest form of control found in modern British poultry units.

Usually a 'black box' is located in the feed room at the end of the house, having dials for setting required temperature and minimum fan speeds, and where necessary other adjustments for heaters and lights.

Timing devices for feeders and egg collection may also be incorporated on the panel.

Inside the proportional control box, the 'solid state' electronics include a 'thyristor', which varies the voltage available to the fan motors, in response to signals from the thermistor, a small sensing device in the house which takes the place of the thermostat. As an alternative to the wall-mounted control box, it is also possible to obtain compact proportional controllers which are suspended in the house.

9 Manure handling

Manure has a value. Most poultry-keepers are aware of the fact; but it is a bit harder to be philosophical when the time for mucking out comes round, which it does with depressing regularity unless you have the land for folds or the other open-floored portable systems described in Chapter 7. To the domestic poultry-keeper, in fact, the value is likely to be mainly in improved yields from the vegetable plot or prize blooms from the flower-bed, but even here a constructive approach will pay off by making the manure easier to handle and more effective in action. For the producer with 100,000 layers, who has to dispose of over 5,000 tons of the stuff every year, the need for efficiency, and the potential return, is obviously that much greater.

It probably will not be any special surprise to anybody who has scraped a droppings board, but birds actually excrete a greater weight of material than they take in as food each day, because their droppings combine urates—the white portion—with faeces, all at a moisture content of 70% or more. This has two effects; it increases, relatively speaking, the amount of material the poultryman has to deal with from his livestock and through the presence of uric acid it produces a manure with a much higher nitrogen content than ordinary farmyard manure (FYM). The produce is richer in two other important plant nutrients, phosphate and potash, as can be seen from the table, based on British Ministry of Agriculture estimates.

Available nutrients in undiluted slurry

Slurry	NITROGEN kg/m^3	PHOSPHATE kg/m^3	POTASH kg/m^3
POULTRY	9.0	5.5	5.5
CATTLE	2.5	1.0	4.5
PIGS	4.0	2.0	2.7

But nitrogen is notoriously volatile and escapes, partly as a constituent of ammonia, when manure breaks down—a process which can be accelerated by bad handling or storage. One frequently recommended way round this is to add superphosphate, which fixes the nitrogen by reacting chemically with the ammonium carbonate forming in the manure. In this way the nitrogen is retained for gradual release in the soil and the 'scorching' effect often observed on grass that has been treated with raw poultry manure is offset. At the same time, smell is reduced and the manure becomes friable and easier to handle.

Superphosphate is easily obtained from garden merchants and can be sprinkled on droppings boards at daily or weekly clean-down times, or into manure pits and under verandas twice a week. Calculate on approximately 0.35 kg ($\frac{3}{4}$ lb)/50 birds/day, equivalent to 2.40 kg ($5\frac{1}{4}$ lb) a week.

Although superphosphate is an undoubted aid where manure is being used on crops, it is also an added cost factor which many poultrymen prefer to sidestep. A simple method of drying on slats, described later in this chapter, is a possible alternative.

When it comes to application as a fertilizer, 'little and often' is the maxim for poultry manure. In the garden it makes an excellent composting material and can be turned into the soil during the spring, at about 113 gm/sq m (4 oz/sq yd) with notable advantage to roses and benefit to other flowers, root crops and greens. It works well as a top dressing for grass and cereals in early spring. Broad recommendations for rates of application on agricultural land appear in the table.

Before seeding

	Grass (tons/acre tonnes/ha)	Roots and Greens (tons/acre tonnes/ha)	Cereals (tons/acre tonnes/ha)
RAW BATTERY MANURE	3.0	3.75	1.75
SLURRY	6.0	6.0	6.0
DEEP LITTER	2.5	3.0	1.5
BROILER LITTER	1.25	1.5	0.75

Manure in the quantities required on farmland almost invariably comes from intensive units. It may be allowed to accumulate in pits below or adjacent to the poultry houses, or it may be augered direct to the muck spreader at cleaning out time. Spreaders come in various shapes and sizes, from tanker type vehicles to tractor-towed trailers working from the power take-off. Depending on their design, they may discharge their load from the side or from the rear. A good spreader will cope with both litter manure and wet slurry, a 5.35 m³ (7 cu yd) model taking about $2\frac{1}{2}$ tons of manure or 2,727 l (600 gal) of slurry.

Occasionally poultry houses are situated on or near arable land. Then it is possible to install a system of slurry irrigation, with pipelines to carry the manure, diluted 3 : 1 with water, to the fields via special spreader nozzles. If slurry is distributed by tanker the dilution rate is 45 litres of water to every 50 kg of manure (10 gal/cwt).

The drawback to the traditional method of disposing of poultry manure, however, is the difficulty of obtaining a realistic return on it.

In some cases, poultrymen are even expected to pay to have it carted away, despite the fact that when fresh it is richer than FYM in several respects. Prejudice among general farmers tends to rest on the fact that it is a volatile substance, capable of 'scorching' if over-applied, yet lacking the long-term beneficial effects of FYM. This rather convenient argument for low prices, coupled with the need to deal with increasing quantities of manure from various production systems has resulted in a mass of ideas for turning it into a profitable by-product. It has been enhanced as a compost, saleable at good prices to gardeners; dried into a feed ingredient for cattle and poultry; processed to produce methane gas. Not a few poultrymen have been glad of the income from manure to keep them solvent when egg prices are bad. There have even been half-serious suggestions that birds should be selected on the basis of their manure-producing, rather than egg-producing abilities.

The method and extent of manure handling naturally varies with the type of stock, and these can be taken in order.

Battery layers

Cages are either arranged over a pit or channel in the house floor, to which the droppings can fall direct, or they have a conveyor belt or scraper between each tier to shift the muck to the end of the cage-bank, where it generally falls into a channel in the floor containing an auger, which will transfer it from the house to a storage tank or waiting spreader

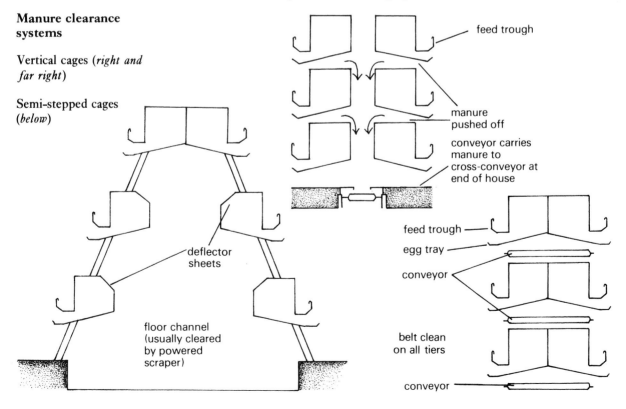

Manure clearance systems

Vertical cages (*right and far right*)

Semi-stepped cages (*below*)

feed trough

manure pushed off

conveyor carries manure to cross-conveyor at end of house

deflector sheets

floor channel (usually cleared by powered scraper)

feed trough

egg tray

conveyor

belt clean on all tiers

conveyor

Droppings from belt-clean cages fall to a cross-conveyor carrying muck out of the house

outside. With belt and scraper systems, the limited space between tiers and the sheer weight of droppings building up demand regular and frequent clearance. Winding may be done manually, but is commonly a job for an electric motor. It is possible to get a portable electric winder which can be carried from house to house.

Since the moisture content and quantity of droppings produced by a flock can fluctuate according to house temperature, food intake, energy in the ration and rate of egg production, to say nothing of the difference between light and heavy strains, estimates of manure output are necessarily broad averages, but in round figures, 1,000 layers can be expected to excrete about a ton of fresh droppings in a week; a ton a day from a house holding 7,000 layers, therefore, and if this has a normal moisture content of just over 70% it will occupy a volume of 1 m³ (35.3 ft³). As it dries the bulk shrinks until at 50–60% moisture content it occupies 0.9 m³ (31.8 ft³) but beyond that, weight rather than volume is lost. At 10% moisture the dry powdery material still occupies approximately 0.9 m³ but weighs just over ¼ ton.

Slurry systems call for the addition of water, of course, and the storage tank must take account of this. Dilution rates of 15% solids are usually applied, which means adding 900 litres of water to each tonne of manure (200 gal to 1 ton) roughly doubling its volume. A week's output from 1,000 birds being 1 m³ therefore, the capacity of the tank must be 2 m³ (70 ft³)/1,000 birds for every week of storage.

Above-ground tanks are frequently circular, and built up in metal plates. Underground holders must be substantially lined to withstand inward pressures, and waterproofed. They must also have a solid cover, and/or a guard fence, for the safety of children particularly.

Some form of agitation is needed with slurry tanks, to keep the solids in suspension so that they can be easily pumped out. The contents of tanks up to about 27,000 l (6,000 gal) can be stirred up by exhaust air from the tanker vacuum pump, when the tanker arrives to collect the slurry. For larger tanks, paddle systems are usually fitted, which can operate at intermediate times, and for up to an hour before the tank is emptied.

The slurry system is not to be confused with lagoons, however. These consist of artificial pools, either inside or outside the house, designed to take the droppings and reduce them by bacterial action. Given the right conditions, primarily a good 'fermentation' temperature and adequate light, lagoons can work quite well in digesting manure by the action of aerobic bacteria without objectionable smells, but if they become unbalanced, allowing anaerobic organisms to come to the fore, smell problems will arise.

An indoor lagoon 1 m (3 ft) deep, filling the area below the birds, is sufficient to take the droppings of a flock stocked at 0.09 m² (1 ft²) of floor space per bird. The allowance for external lagoons is 0.8 m² (1 yd²) of surface area per bird.

Lagoons offer the advantage that they can be left to their own devices, for years if necessary, without manure removal, but they have been more successful in those parts of America where warmth and good lighting are available than in England, where winter lighting particularly tends to retard the breakdown of droppings.

Deep pit

For trouble-free handling of manure from battery cages, the deep-pit system is highly popular among British egg producers. 'Pit' is perhaps a misnomer, since in most cases it is a walled-in area above ground, 2 m (6½ ft) high, on which sits the production unit, with its cages and services.

Basically, droppings from the caged birds find their way to the pit

Diagrammatic representation of the deep pit with in-pit slats system. The drawing is not to scale and is shown purely to illustrate the method. Differing quantities of manure on the slats and on the floor may occur in practice. Commercial and ADAS advice is available to producers considering this type of housing.

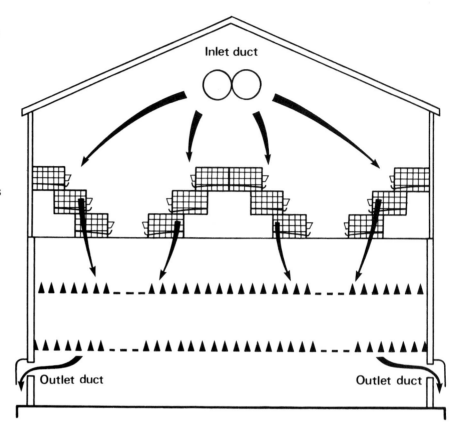

floor, where they are allowed to accumulate to the end of the flock's life and sometimes for several flocks before removal.

Pressurized ventilation, drawing air in at the ridge, driving it down through the cages and out via outlets in the pit walls, has a drying effect on the manure while at the same time maintaining low humidity levels in the house. Doors at the end of the pit allow access for a tractor with a front-end loader when the time for removal arrives.

A refinement of this system, worked out by Arnold Elson and Alec King, specialists in the Ministry's Agricultural Development and Advisory Service (ADAS), is slat drying. Wooden, slatted frameworks are suspended below the cages at two levels. The upper slats, 15.2 cm (6 in) wide and with the same width between them, are situated 76 cm (2 ft 6 in) below the cages. Droppings that miss them and pass through the gaps, fall to the second slatted framework a further 76 cm (2 ft 6 in) below, made up of 7.6 cm (3 in) slats spaced the same distance apart.

Cones of manure are allowed to build up on the frameworks where drying is rapid and thorough. By the end of the laying season, the manure cones have a 15% moisture content—equivalent to mechanically dried material—yet retain much of their original value when fresh. Analysis at the Ministry's Experimental Poultry Station, Gleadthorpe, where a prototype unit was set up, has shown typical samples to contain 4.9% nitrogen, 2.1% phosphate, 2.3% potassium and 5.2% calcium.

Hinged to the house stanchions in the pit, and supported at their outer edges by ropes, the frameworks are dropped to hang vertically, releasing the manure on to the floor when ready for removal. At this stage it will be mixed with the 15–20% of droppings which have reached floor level. Unless special precautions, in the form of a third row of slats or a layer of straw, have been taken, the floor droppings will be at a higher moisture content of 40–60%, raising the mean moisture level to 25–30%.

It is hard to fault this method of manure handling for any deep pit system employing Californian or flat-deck cage layouts with manure falling to the pit over a wide area. Installation and labour costs are low and the product is of a high grade for milling as livestock feed or use as fertilizer. It may be necessary in humid climatic conditions to supplement with in-pit fans. Vertical tier cages, which are being increasingly adopted because of their optimum use of house space, expose its limitation however. They provide no direct way for manure to fall to a pit. The specialists believe a slot system in the floor, through which manure can be directed so that it again forms cones in the pit, may be the answer here.

Deep litter (layers)

Deep litter houses for layers or breeding stock frequently employ a droppings pit under a slatted roost. During a year-long season about two thirds of the droppings will be excreted into the pit, amounting to 35 tons at a moisture content of 70% from 1,000 birds. As this dries to around 50% moisture, its weight will reduce to about 20 tons. The balance of the manure, some 18 tons, becoming mixed with the litter on the open floor, will come down to a moisture level of about 30% to give it a final weight of 6–7 tons. In all, therefore, approximately 27 tons of manure, plus a further 3 tons or so of litter will require to be moved after a year's production with 1,000 layers.

Deep litter (broilers)

By the nature of their production broilers present a different pattern of manure management from layers, whether on litter or in batteries. The growing birds produce smaller quantities of droppings, but their short, seven- to eight-week cycle means that house clean-out is a regular and frequent requirement.

The output per 1,000 broilers in eight weeks can be estimated at about 4.3 tons of fresh droppings. Under controlled environment, if ventilation, control of drinkers and general litter management has been good, moisture loss will reduce the weight of manure to around 1.7 tons. But this, of course, is mingled with the litter, placed in the house at the rate of 550 kg per 100 sq m (0.5 tons per 1,000 ft²), equivalent to 0.32 tons per 1,000 birds stocked at 0.65 sq ft. So a little over 2 tons of broiler litter per 1,000 birds must be removed at the end of each cycle.

Broiler breeders

Where birds are kept for longer periods on litter, such as broiler breeders, manure accumulation is greater, but otherwise follows the same trends to moisture loss as other floor systems. A normal stocking period will be thirty-eight weeks, during which these heavier, meat-type birds will excrete some 42 tons of droppings at the 70% + moisture level. This will reduce by evaporation to something under 50%, giving a manure weight of around 16 tons, and a total weight, with litter, of some 19 tons from each 1,000 birds to be shifted.

Turkeys and waterfowl

Output of manure from turkeys on deep litter will amount to 10–12 tons per 1,000 birds (including the weight of the litter material) after fifteen weeks. After twenty-four weeks the accumulation will have reached some 25 tons. It can be handled and treated much as chicken manure.

Ducks produce wetter droppings than chickens, which might sometimes pose difficulties in battery brooding, but in general their manure has similar nitrogen, phosphate and potash levels to other poultry, making it a useful plant food. To prevent leaching out of its nutrient contents by rain, in common with other forms of manure, it should be stored under cover, and then only briefly. Slurry systems and normal handling methods are all suitable for duck manure.

Geese have a copious output of wet manure, of lower value than the other species as a plant nutrient, but as they are more generally kept on free range, disposal is not usually a problem. Where any intensive system is used with these birds, a slurry system would appear to offer the most convenient handling method.

Waterfowl, in fact, offer another option for droppings. In Eastern and some central European countries, they have access to pools containing carp, which benefit from the nutrients entering the pool, and provide the duck or goose farmer with a second source of income.

Mechanical drying

The possibility of turning poultry manure into a profitable commodity by mechanical drying is no myth, but this is an investment that does need to be approached with caution. Some producers with large flocks have found that the drier was a better source of income than their eggs.

The attraction of the process is that it produces a granular substance, pleasant to handle, with no objectionable smell, which can be bagged off in, say, 12 kg (26$\frac{1}{2}$ lb) polythene or paper sacks as a fertilizer, for sale through garden equipment stores and horticultural suppliers at far higher unit prices than could be obtained for the bulk wet manure. Alternatively—and potentially with more profit—it can be used as a food ingredient for ruminants or chickens.

No system has yet been invented that totally eliminates 'nuisance' smells from mechanical driers, except at high cost. Chemical masking agents, air scrubbers and Dutch drains (in which exhaust gases are entrained through a filter-bed of stones) have all been employed. Filtration systems work, but not totally, and involve a certain amount of management. Masking does not answer the neighbour's objections to any sort of smell wafting over. After-burners have perhaps come closest to the ideal. These bring the exhaust gases to a very high temperature before releasing them to the atmosphere, but they carry a penalty in raising fuel costs.

Local objections to smells are probably the biggest single enemy to driers, and any new installations must be sited with care, studying prevailing wind patterns and the distance from possible objectors, but depreciation, running costs and availability of markets are also important factors.

Mechanical driers are generally oil or gas-fired and either work on a continuous or batch drying process. The continuous machines are the more common and most models consist of a rotating, horizontal cylinder into which the wet manure is fed steadily by an elevator from a holding pit and subjected to heat as it is passaged through various chambers, to emerge in a dried form. Batch driers may be relatively simple flat-bed arrangements on which the manure is spread and heated from above, or, more usually now, they are large, drum-shaped machines, enclosed at the top, in which the manure is heated and agitated by paddles. Whatever the design, these mechanical driers all operate at high temperatures with a corrosive substance. As a result they frequently incorporate stainless steel and high grade materials in their construction, which sets their initial cost in the thousands—more often the tens of thousands—of pounds, while the necessarily hard mechanical life means that the prudent producer will write off the investment in five years if possible.

Unquestionably, the mechanical drier is of most potential use to the large-scale egg producer in quickly reducing a sticky problem to manageable proportions. The standard 1 ton of wet manure from 1,000 battery hens a week will dry down to 0.28 tons of material at around 10% moisture content. By this estimate, the 20 tons from a 20,000 bird flock will reduce to 5.6 tons, the 50 tons from 50,000 birds to 14 tons and the 100 tons from 100,000 birds to around 28 tons.

Even with a drier, though, it is still in the interests of the producer to keep the water content of the raw manure as low as possible, by paying attention to leaky drinker lines, wet storage conditions and so on. Twice as much water must be evaporated from manure at 80% moisture content as from that at 65%, and that means twice as much fuel. Likewise, fuel costs increase steeply with the use of an after-burner to control smell. Where neighbours are close and maximum reduction in smell is essential, fuel for the afterburner can equal the amount needed for the main drying operation.

A drier operating at 80% efficiency, which is an accepted norm, takes 1 litre of oil to evaporate 9 kg of water (equivalent to 1 gal of oil to 90 lb water) on the basis of scientific estimates that 4.55 littres (1 gal) of oil evaporates 50 kg (1 cwt) of water at 100% efficiency. To reduce 1 ton of manure at around 75% moisture content to 0.28 tons at 10% moisture, therefore takes just under 82 litres (18 gal) of fuel for drying alone. Running costs can be calculated on this level. On top of this, maintenance, labour, electricity and packaging for the end product will add their bit, of course.

One of the difficulties in marketing DPM is its inherent variability. Nitrogen level, for instance, is higher if the manure comes from birds on a high energy diet and other flock differences can make wide variations in the qualities of DPM for either fertilizer or feed. Although some of these can be ironed out by consistent management, it could be difficult to put a guaranteed analysis of contents on each bag of DPM sold as fertilizer, to meet the requirements of the Feeding and Fertilizer Act. There is nothing to prevent it being sold as a soil conditioner, however.

Used as an animal nutrient, DPM is either sold to feed compounders or re-cycled as a proportion of the producer's own livestock rations. Being high in fibre and crude protein it is well suited to the feeding of beef, sheep and dairy animals. For poultry it cannot be considered so ideal. It is low in energy for instance and unsuitable for broilers. As a replacement for expensive cereals in layer rations, however, it has proved a successful cost saver at inclusion rates of 10–12%.

One farmer who has both beef cattle and 150,000 layers and runs two driers, uses the following proportions:

		%
For beef:	Vitamin/mineral balance	5
	Barley	60
	Maize (corn)	15
	DPM	20
For layers:	Protein/vitamin/mineral concentrate	15
	Wheat	13.3
	Maize	43.3
	Barley	13.3
	Limestone	5
	DPM	10

Producers in America tend to favour DPM for replacement pullets rather than layers. They also combine manure, as it comes from the poultry house, with chopped corn fodder or chopped hay and put the mixture into silos to be later fed to cattle.

Understandably, safeguards are necessary to ensure that pathogens, heavy metals and/or drug residues do not build up in DPM. The EEC Directive 74/63 which governs maximum permitted levels of undesirable substances in feed ingredients, stipulates that no more than two parts per million of arsenic; 5 ppm of lead; 0.1 ppm mercury and 0.01–0.02 ppm aflatoxin B must occur in compound feeds. Specifically in the UK the Agriculture Act 1970 specifies the general quality requirements for feeds and the Medicines Act 1968 sets out the levels and types of additives for use in animal feeds.

In America, legislation in some States allows the feeding of DPM to food-producing animals. In others it is sanctioned only if fed to pullet replacements, brood cows and replacement heifers. It must be salmonella-free and a licence to process is required. Regular analysis of the product is obligatory. The Food and Drugs Administration also takes a hand in waste recycling legislation. Producers in the US are quite frequently taken to court for spreading manure on land near homes or businesses. In some cases even a mile has not been acceptable. Air discharged from production units and offensive to local residents, is a major concern to poultrymen, taking precedence over all zoning laws.

Drying does not exhaust the possibilities for manure treatment, however. Systems for fast composting have been developed, to produce a relatively dry—17–20% moisture content—material retaining its nutrient value in nitrogen, phosphates and potash. The process consists of a reactor, into which fresh manure, mixed with a proportion of the dried product, is introduced and circulated plus oxygen, over several days, before being discharged as a high value compost. The operation is said to be free of smell and much cheaper than thermal drying, but of course the return on the resultant 'soil improver' will be less than for a feed ingredient.

Methane has long presented a fascinating dream of riches from manure, and there is no doubt that the gas can be recovered from poultry slurry. It is put into a digester tank, heated to 27–37° C (80–100° F) and the gas given off can be collected. A Devonshire engineer, Harold Bate, estimating that 50.80 kg (1 cwt) of manure produces the methane equivalent of 15.15 litres (4 gal) of petrol, has employed a methane plant and invented a carburettor adaptor enabling an ordinary car to be run on the gas, but in practical terms, pressurization costs and disposal of the slurry residue put question marks over the value of methane recovery for the poultry farmer.

The ultimate solution for those with a manure disposal problem may seem to be to 'flush it away' but this calls for careful consultation with local authorities. The activity of decomposing waste is measured by its 'bio-chemical oxygen demand', or BOD which means the amount of oxygen required by aerobic micro-organisms under test conditions to reduce a given quantity of material. It is expressed in mg/litre. Domestic sewage has a BOD of 250 mg/litre and by comparison it would require 36.4 litres (8 gal) of water to be added to each day's droppings from one bird to give an equivalent sewage. The River Thames has a BOD of about 14 mg/litre which is still being reduced and this is about the maximum acceptable level for effluent discharge into rivers or streams in the UK.

10 Food and water

by John Portsmouth

Poultry can survive, grow and produce eggs on an extremely wide variety of food stuffs. You can feed birds on household scraps, and provided you understand the basic principles behind providing a balanced ration, the results achieved will not be disappointing. Larger poultry-keepers however, must use a balanced compound in a form which is perfectly acceptable to the birds and which can be left in front of them throughout the hours of day length. The large poultry enterprise will look towards milling and mixing its own rations through the blending together of purchased protein ingredients such as fish meal and soya bean meal with the grains wheat, barley, and maize or corn. To complete the nutritional blend the minerals, trace elements and vitamins will be added, and the whole will form a balanced ration on a par with a purchased compound which has been designed to produce the most economical output of eggs and meat.

Unlike the domestic poultry-keeper, to whom it matters little whether he takes twelve or twenty weeks to produce a meat bird weighing 4 kg ($8\frac{3}{4}$ lb), the specialist poultry unit depends for its very survival on the production of large numbers of birds in the shortest time; for example, broilers weighing 2 kg ($4\frac{1}{2}$ lb) in forty-nine days. But with both scales of operation feeding stuffs account for the major portion of the cost of production and it is essential that the poultry-keeper has a sound knowledge of the underlying principles of feeding.

The nutrients

Foods are composed of a number of nutrients which can be grouped into protein, carbohydrates, fats, minerals, vitamins and water. According to the particular type of work carried out, all these nutrients must be provided in the correct proportions, an excess of one not compensating for a deficiency of another. The quality of the balance is very important. Not all nutrients are available to the bird, some being in a form that is unusable by the bird despite their apparent high value. The energy requirement is expressed as metabolizable energy or ME (MJ/kg or kcals/kg).

Proteins

Proteins are essential constituents of all tissues. They are composed of complex substances, and the amino acids of one protein are not the same as the amino acids of another one. Therefore, one protein cannot be replaced by another one in order to obtain the same balance of amino acid.

The amino acids are divided into two groups, the essential and the non-essential. Non-essential amino acids are those which the bird can manufacture for itself from the proteins which it is fed, whilst the essential amino acids cannot be manufactured by the bird and therefore have to be provided by the food. Essential ones are arginine, glycine, histidine, leucine, isoleucine, tryptophan, threonine, methionine, lysine, phenylalanine and valine. An excess of amino acid is broken down and used as an energy source. A balanced ration prevents this from occurring. Proteins are very expensive, and energy can be provided much more cheaply from other sources.

The birds' ability to store amino acid is limited. The essential amino acids, in the correct proportions, have to be supplied in the right quantities at the same time. Protein requirement is highest when the bird is growing very rapidly. With young chicks, broilers and turkey poults the protein requirement is at its highest in the first four to six weeks of age, and gradually declines as the growth rate slows relative to the amount of food consumed. The mature chicken which has virtually stopped growing, needs protein to replace worn-out tissues, whilst the laying hen needs protein for egg formation and some small growth. The precise amount of protein and amino acid required in the ration depends on the level of feed intake. When feed intake is high, the percentage of protein in the ration can be less than when the feed intake is low. As feed intake is controlled by the amount of energy in the food it is important to relate the non-energy portions of the ration to the energy portions of the ration. This is known as the energy protein/amino acid ratio, and use of this ratio allows lower levels of nutrients when the energy is low and higher levels of nutrients when the energy is high to compensate for high and low feed intakes respectively.

The breakdown of food protein to respective amino acids and selection of the essential amino acids and synthesis of the non-essential ones in the correct proportion with their final build up into egg protein involves critical amounts of food protein in precise quantities to enable the bird to produce large quantities of eggs of the required weight. The domestic poultry-keeper feeding household scraps and vegetables is most unlikely to achieve this critical balance. Despite this, because he is able to give more individual attention to his small flock, that which is lost is often compensated.

Carbohydrate

Carbohydrates supply energy. Carbohydrates can be divided into sugars, dextrins, starches, celluloses, lignins and pentasans. Poultry have limited ability to digest fibre. The available energy of a poultry ration is increased by reducing feed ingredients containing high levels of crude fibre.

Excess consumption of energy is converted by the body into fat, which can be converted back into carbohydrates for use as work energy. But it can lead to over-fatness and obesity, which is detrimental to egg production and feed efficiency in both laying and fattening birds, and is an undesirable eating characteristic. Over-consumption of calories easily occurs and unless some means of physical restriction is introduced it is difficult to prevent, but provided it is not excessive the effects will not be seriously damaging to productive and economic performance.

Fats and oils
The other main source of energy to the bird is fat and oil. Fat is two and a quarter times as rich in energy, weight for weight, as carbohydrate. In commercial broiler production where energy may be a limiting aspect in formulation a certain amount of fat can be used and it will take up a much smaller amount of space in the diet for a greater output of calories. Fats are made up of glycerol and fatty acid. Certain fatty acids are essential to the bird, namely linoleic and oleic acid, which must be provided in fairly constant amounts in order to obtain optimum productive performance. Meat and fish products are rich in oils, as are the vegetable proteins such as soya bean, groundnut and sunflower meal. However, it is normal practice to extract the oil in processing from the vegetable proteins, so the product which is fed to poultry is normally extremely low in fat and contributes very little to the fatty acid requirement of the bird. On the other hand, meat meal and fish meal still contain relatively high quantities of fatty acid.

It is necessary to incorporate an antioxidant into the ration to prevent oxidative rancidity which imparts an off-flavour to feed. Chickens can tolerate and digest high levels of fat, and provided it can be incorporated without seriously affecting the physical form of the ration when it is in a meal form or when it is pelleted, then levels up to approximately 5% are a practical proposition. Above this difficulties arise with feed manufacture. Domestic poultry-keepers should ensure that table scraps are used as quickly as possible and not left to deteriorate through rancidity. Rancidity will cause digestive disorders, diarrhoea and a refusal to eat. Loss of both physical and economical performance will therefore result. Normally the essential fatty acid requirement for both laying and fattening birds is adequately covered through the natural ingredients in the ration. Occasionally, however, beneficial effects can be obtained by adding linoleic acid through the use of either soya bean oil or maize meal oil which are rich sources of this fatty acid.

Minerals
Minerals are extremely important for many bodily functions, but perhaps the most important role is with the skeleton. The majority of

calcium and phosphorus is found in the bird's skeleton, whilst potassium is found mainly in the muscles, iron in the blood and iodine in the thyroid gland. The requirements depend upon the age of the bird and the production in which it is involved. For example, the laying bird's requirement for calcium is extremely high compared with the growing bird, because the egg shell is composed almost entirely of calcium carbonate.

Calcium

Whilst all food contains calcium, those which are most rich in this mineral are meat and bone meal, fish meal and of course, products such as limestone, oyster shell and dicalcium phosphate. The calcium level of vegetable material such as the cereal grains and vegetable proteins is extremely low relative to the laying bird's requirement for calcium. The growing bird needs calcium to develop a strong skeletal system capable of carrying large amounts of protein tissue. The laying bird, in addition to this requirement, needs to have sufficient calcium to produce an egg shell a day. In practice, an egg every day is rarely obtained, but from the point of view of determining the bird's requirement it is necessary to consider the ultimate objective. Each 58 g (2 oz) egg contains approximately 2 g ($\frac{1}{20}$ oz) of calcium but because the laying bird is only able to retain approximately half of the calcium it consumes, it is necessary for the ration to contain some 4 g ($\frac{1}{10}$ oz) of calcium. A laying bird consuming 100 g (3$\frac{1}{2}$ oz) of a compounded ration each day needs 4% of that ration made up of calcium. If the bird consumes 150 g (5$\frac{1}{4}$ oz) a day, although it still needs 4 g ($\frac{1}{10}$ oz) of calcium, the percentage calcium in the ration is reduced to 2.7%. Calcium for the maintenance needs of tissues is very small and is adequately catered for from a level of around 1% calcium of the ration.

In the absence of calcium, both the skeleton and egg shell will suffer. Leg weaknesses will occur and growth will be impeded in young birds. In laying birds, shells become weak, thin and fragile. In an effort to compensate for a lack of calcium the bird will draw on this mineral from its skeleton with the result that it is unable to stand due to leg weakness. In cases of continued deficiency, death will occur. Calcium is closely linked with both phosphorus and vitamin D_3. Vitamin D_3 is necessary for the transport of calcium across the bird's intestinal wall. Phosphorus, on the other hand, is closely linked with calcium in the skeleton and it is therefore essential that the correct proportions of both calcium and phosphorus are provided in the diet. For growing birds the ratio of calcium to phosphorus is in the order of two parts calcium to one part of phosphorus, whilst with the laying bird the ratio is approximately six parts of calcium to one part of phosphorus. Without adequate levels of vitamin D_3 the calcium deposition will be impaired and the condition known as rickets will occur. The domestic poultry-keeper should ensure that the calcium requirement is covered by provision of a free-choice supply of either limestone grit or oyster shell grit. This will balance up any deficiencies in the household scraps and vegetables. The birds' calcium requirement is provided in compounded feeding stuffs and there is generally no need for further addition. The integrated organization, milling and mixing its own feed, supplies calcium through the natural ingredients and by limestone flour and in order to reduce dustiness which occurs when limestone flour is used it may be necessary to use a pin head size limestone grit.

Excess calcium should also be avoided. Since excess is excreted, it has the ability to tie itself to certain essential trace elements, which if not included in the right quantities in the ration will lead to various nutritional disorders.

Phosphorus
The laying ration should contain in the order of 0.65% total phosphorus. Growing birds, either for replacement egg production or for meat, should have a phosphorus level of approximately 0.7–0.75%. Not all the phosphorus supplied in the diet is available to the bird. Young birds are unable to use a large proportion of the phosphorus supplied by vegetable or plant materials, known as phytin phosphorus. Adult birds can cope with greater quantities of this phytin phosphorus. Rich sources of phosphorus are meat and bone meal, fish meal, and other animal products. Additionally, phosphorus is supplied through sources such as dicalcium phosphate. Any inadequacies of the diet are usually made up through the addition of dicalcium phosphate or similar phosphate sources. The available phosphate level of a ration is more important than the total phosphate level. Young birds' rations should contain 0.33% available phosphorus for layers and 0.45% available phosphorus for the meat birds.

Salt
Salt is composed of sodium and chloride and both of these elements are very important for egg production and growth. To cover the requirement of salt most rations are supplemented with 1–2 kg ($2\frac{1}{4}$–$4\frac{1}{2}$ lb) of salt. As a percentage of the ration the salt content should be in the region of 0.4% for all classes of stock. For laying birds and broilers, of the 0.4% salt, some 0.12–0.15% should be as sodium. A lesser amount than this is adequate for replacement growers, whilst 0.14–0.15% is required by the laying bird.

In the absence of salt birds are extremely nervous, will not grow well and egg production will suffer. When excess salt is supplied water consumption increases dramatically with the result that droppings become extremely wet causing impairment to the bird's digestive system and problems where birds are kept on litter. Meat and bone meals and fish meals are rich sources of natural salt.

Trace elements
Trace elements can be defined as small quantities of essential minerals

which it is usually necessary to add to the diet at levels in the order of less than 100 ppm or 100 g ($3\frac{1}{2}$ oz) to the tonne of feed. Despite their extremely small requirement, they are very important to the well-being of the bird and its productive, economic efficiency. Two of the most important trace elements are manganese and zinc. Manganese is most well-known for its association with the condition known as perosis in growing birds (see next chapter) and chondrodistrophy in unhatched eggs. Practical rations will provide in the order of 25–30 ppm of manganese from natural sources, and in non-competitive situations where birds are not made to grow fast or to lay large numbers of eggs, this amount is usually adequate to prevent nutritional disease. Manganese should be provided as a supplement to the diet at between 50 and 100 ppm, depending upon the class of stock. For broilers and breeding birds the supplemental rate should be 100 ppm, whilst for growing birds 50 ppm is adequate. Turkeys, ducks and geese also need adequate manganese in their diet and it is usually considered necessary to supply in the order of 70-80 ppm of supplemental manganese for ducks and geese, whilst a level in the region of 100 ppm should be added to turkey growing and breeding rations.

Zinc

Adequate levels of dietary zinc are important for all classes of poultry, including ducks, turkeys and geese, and most trace element supplements which are added to poultry rations should provide in the order of 40–70 ppm of zinc depending on the class of stock.

Other trace elements

Iron and copper are involved in the production of red blood cells. They assume greater importance after bleeding has occurred following an outbreak of coccidiosis. A deficiency produces anaemia and in extreme cases, depigmentation of feathers. The iron content of feed stuffs is normally adequate to meet the nutritional requirements.

A deficiency of copper leads to anaemia, depressed growth and bone disorders. Copper at high levels has a therapeutic and growth stimulating effect; it has been used for many years to treat conditions such as gizzard erosion, enteritis and mycoses, and levels of between 150–250 ppm have been shown to enhance growth rate and feed efficiency. Copper is usually included in mineral supplements as a safety margin to the natural feeding stuffs.

The primary role of selenium is in the prevention of the condition known as exudative diathises in chicks. This element is intimately tied to vitamin E nutrition, although both nutrients play very specific roles in poultry nutrition. Selenium responsive diseases can be cured through the use of vitamin E, but the amount of selenium needed to prevent those specific conditions is much less than that needed by vitamin E. Selenium should be added to all rations at a level of around 0.1 ppm.

A deficiency of iodine reduces the bird's metabolic rate. Iodine controls the rate of release of the thyroid stimulating hormone. When iodine is deficient in the ration an enlargement of the thyroid gland, called goitre, occurs. A deficiency reduces hatchability and growth rate. Mineral supplements contain small amounts of iodine, and iodized salt is a convenient way to add it to the diet.

Molybdenum deficiency causes a feather abnormality and retarded growth in chicks. It is necessary for good hatchability and although many of the ingredients used in poultry feeding stuffs contain molybdenum, supplemental molybdenum may be added at 1 ppm.

Vitamins

A vitamin is an organic compound which is distinct from carbohydrate, fat, protein and water, and is present in foods in very small amounts. It is necessary for the development of normal tissue and for health, growth and maintenance of that tissue, and when absent from the diet or incorrectly absorbed or utilized, results in a specific deficiency. The following vitamins are added to some or all of specific poultry rations: vitamin A, vitamin D_3, vitamin E, vitamin K, vitamin B_1, vitamin B_2, vitamin B_{12}, vitamin B_6, nicotinic acid, pantothenic acid, biotin, folic acid, choline and vitamin C.

The vitamins are divided into two groups depending upon their solubility in either fat or water. Fat soluble vitamins are A, D, E and K and are found in feed stuffs in association with fat; the remainder are water soluble. They are not associated with oil and an alteration in the oil content of the food does not affect their absorption. Fat soluble vitamins can be stored in appreciable amounts in the body, whilst, except for vitamin B_{12}, the water soluble vitamins are not stored and excesses are rapidly excreted. It is essential that a continuous supply of water soluble, B group vitamins are provided in the diet in order to avoid deficiencies occurring, and to maintain adequate growth and feed conversion.

The chicken has little or no ability to carry out microbial synthesis of vitamins in the intestinal tract. Indeed, the reverse is true in that intestinal micro-organisms in the chicken tend to compete with the bird for dietary vitamins and therefore the chicken has a high requirement of vitamins. The recommended levels of vitamins in practical diets are shown in the table on pp. 144-5. The levels given are high enough to take care of the fluctuation in environmental temperature, energy content of the diet, and other factors which may influence the feed consumption or which may influence the vitamin requirements. They are not those levels, however, which are recommended for use in hot and humid climates of the world. Under those situations the allowances or recommendations are by necessity increased considerably. Some results of lack of vitamins can be seen in Chapter 11.

Water

Water is one of the most important inorganic compounds in the bird's body. It is essential for all normal functionings and is the basic substance of the blood, the cells and the intercellular fluids acting in the transport of food nutrients, metabolites and the transport of waste products to and from all the cells of the body.

Without feed chickens can live for a considerable amount of time. Without water, even for one day, physiological changes occur, resulting in a marked reduction in growth rate and egg production. Laying birds may react by moulting all their feathers. The water content of the chicken is high in the early periods of growth, 80–85%, and gradually reduces as the bird gets older to around 60%. Water content of the whole egg is about 65%, and its importance in maintaining body growth in egg production is obvious. The bird can satisfy its requirement by drinking water which has been provided for it, or through the ingestion of green foods such as cabbage and root crops although this will not support good growth or egg production. Birds which have access to fresh green foods will drink less water than those which do not.

Temperature has a marked effect upon water consumption. Water consumption may increase by 100% for a rise in the environmental temperature from $21°\,C\,(70°\,F)$ to $32°\,C\,(90°\,F)$. In hot climates water consumption will be higher than it is in cooler climates, and for birds housed in artificial temperatures of around $21-27°\,C\,(70-80°\,F)$, water consumption will also be higher. Warm and very cold water will be rejected by the chicken. Adult laying birds consume between 18–23 litres (4–5 gal) of water per 100 birds per day. This figure should be used as a guide. The birds must have a clean supply of freely available water at all times. Apart from layers, broilers consume 19.6–21.9 litres ($4\frac{1}{4}$–$4\frac{3}{8}$ gal) at seven to eight weeks and turkeys 42.8 litres ($9\frac{1}{2}$ gal) from fifteen weeks onwards, but it is important to provide water in excess of what any birds need, in containers which allow many birds to drink at the same time. With both ducks and geese water should be offered in containers which allow the birds to immerse their heads fully as it has been found that these species of poultry need to lubricate the inner eye membrane by continual head dipping.

Non-nutrient food additives

Certain non-nutritional feed additives enhance growth and egg production. Those involved in growth are known as chemical growth promoters. Most commonly used products are Zinc Bacitracin, Nitrovin, Virginiamycin and Arsanilic Acid. Chemical products used for both growth promotion and egg production have to be approved by the Ministry of Agriculture, Fisheries and Food. As a 'rule of thumb' guide most growth promoters are capable of improving growth rate in the order of 2–3% and improving feed conversion by 2–$2\frac{1}{2}$%.

The cost of these substances varies considerably, and normally the most effective is the one which costs least. Zinc Bacitracin for use as both a growth promoter and egg promoter is one of the most popular chemical compounds on the market. For egg production, the use of Zinc

Bacitracin between 80 g and 100 g per tonne ($2\frac{3}{4}$ oz and $3\frac{1}{2}$ oz per ton) of feed increases rate of lay in birds with an average egg production of 250 eggs by 3–4%, whilst birds with higher flock averages increase their rate of lay by 2–3%.

Yolk pigments

A rich, golden yolk is a necessary characteristic of eggs sold into the UK market. Good yolk pigmentation can be obtained through the use of natural ingredients such as maize meal and grass meal. These ingredients supply xanthophylls, which impart yellow-orange colour to yolks, but natural ingredients vary considerably in the amount of pigment they provide. It is therefore necessary to supplement the natural pigments with synthetic xanthophylls which can be obtained in yellow and red materials, a blend of the two resulting in an orange pigment. Depending upon the depth of colour required in the yolk, the ration can be topped up with the required amounts of red (canthaxanthin). The pigmenting power of the canthaxanthin is some four and a half times greater than that of the pigments found in maize and grass meal. Domestic poultry-keepers obtain satisfactory yolk colour through the provision of fresh green foods, these being rich sources of natural pigment.

American producers prefer a medium to light coloured yolk because eggs with dark yolks are more difficult to candle on mass candling equipment. It is also difficult to obtain evenly matched yolks as you move to the darker shades, because the rate at which a hen is laying has a slight bearing on the amount of xanthophyll available to colour each of her yolks.

Feeding stuffs commonly used in the manufacture of poultry feeds

The accompanying tables show a comprehensive list of all types of ingredients which are used in the manufacture of the many types of poultry feeds. Tables relate to the average composition of the feeds, and therefore there will be differences in individual values depending upon the season, variety of product growth, and in the case of processed foods, the way in which they have been prepared (temperature treatment in the case of soya bean and fish meal). As a guide to the smaller domestic poultry-keeper using home-grown vegetables and scraps from the house the following notes will be of use.

Bread
Many domestic poultry-keepers include bread in the diet of the chicken. The moisture content, however, varies from 44% to 9% in bread which has been artificially dried. On a dry matter basis, the bread contains around 14% protein and 80% carbohydrate, this being largely in the

form of starch. It can form a very useful part of the ration and it is best used when it has been dried and crumbled, and mixed with other materials.

Grass
Grass is a natural rich source of vitamin A and yolk pigmenting materials. Unfortunately, it also contains fibre which is not used by the bird. If dried grass meal is included it should not constitute more than 5% of the ration. Fresh grass can be given to poultry, although in this form its useful value is quite low and it should be regarded more as a source of yolk pigmenting material.

Potatoes
Raw potatoes are very indigestible and should not be given to poultry. Potatoes should be cooked to break down the starches into readily available carbohydrates. When they have been cooked, rations with up to 50% of potatoes have been successfully fed to laying birds and up to 40% for growers. They are not recommended in amounts of more than 10% of the diet for young chicks. Potato peelings can also be given, and should also be boiled first.

Ration specifications

Ration specifications are an interpretation of the chicken's nutrient requirements and presented in such a way that the ration formulator has only to follow the recommendations in order to ensure that the required nutritional balance is achieved. The following notes should be read in conjunction with the tables at the end of this chapter.

Replacement pullets
The chick ration is designed for feeding to pullet chicks being reared as replacement egg layers and should be fed from day old to between six and eight weeks of age. No other food should be fed. This specification is also recommended for chicks being reared as replacement breeding stock, that is, commercial breeders and broiler breeders. It should be fed *ad lib*.

Growers' rations
Grower No. 1 is fed from six to eight weeks of age until approximately two weeks before egg production begins. It can be fed *ad lib*, but is more specifically designed for feeding in controlled feeding pullet programmes on which specific advice can be obtained from the relative chick suppliers. Grower No. 2 has a lower nutrient density than Grower No. 1 and is recommended for feeding as the second stage of a two stage replacement feeding programme. It should be introduced at twelve weeks of age for light bodyweight hybrids and fourteen weeks of age with medium/heavy hybrids. Restricted feeding programmes aim at

controlling the energy intake of growing birds in order to prevent them laying down surplus body fat. Restricted feeding programmes generally lead to greater uniformity of the birds at point of lay and to lower feeding costs because the feed intake is restricted. It is important either to turn to *ad lib* feeding or to markedly increase the feeding rate should an outbreak of disease seriously impair the growth of the growing pullet.

Laying rations
Laying foods should be introduced some two weeks before egg production is due to commence so that the rapidly developing pullet has the opportunity to lay down reserves of nutrients, especially calcium, in readiness for egg production. When the bird begins to lay, small quantities of calcium are removed from the skeleton and used for shell formation. Should the so-called buffer store of calcium be inadequate a drain on the skeleton's calcium reserve may be excessive.

Laying birds eat sufficient food to cover their energy requirement. Three layer ration specifications are provided so that laying birds with small, medium or relatively large appetites can be provided with sufficient nutrients which will result in a high level of egg production being maintained with eggs of average to good size.

Breeding rations
In terms of protein, amino acids and energy the requirements of breeding birds are the same as commercial egg layers. The one major and essential difference between these two classes of stock is in the vitamin and trace element requirements. Two ration specifications are shown with the Breeder No. 1 being recommended for the first forty weeks of lay, Breeder No. 2 following this ration to the completion of the breeding period. This two-tier programme is particularly useful for broiler breeding birds where the level of feed restriction is particularly tight in the periods just prior to lay and in the early weeks of lay.

Broiler rations
The nutrient specifications shown relate to flocks of hatched broilers which are killed between forty-five and fifty-six days of age. As-hatched flocks of broilers should weigh approximately 1.75 kg (4 lb) at forty-five days and have a FCR of 1.95:1. Broilers of forty-eight to forty-nine days of age should weigh 1.8 kg (4¼ lb) with a FCR of 2.0:1. At fifty-three days, the birds will weigh approximately 2.05 kg (4½ lb) with a FCR of 2.10–2.15:1. These weights and food conversion rates can be obtained with good management and birds housed at stocking densities of 0.45 sq m–0.50 sq m (⅓–½ sq ft) per bird.

The feeding programme used in broiler production has a marked effect upon performance. There is no room whatsoever for even the smallest margin of error. Growth rate must start and continue unchecked. A typical feed programme is 1,000 kg (20 cwt) per 1,000 birds of starter feed and 1,500 kg (30 cwt) per 1,000 birds of the growing feed.

A mill and mix unit on a large farm

For birds killed at between fifty-two and fifty-six days, the starter should be fed at a rate of 750 kg (15 cwt), the grower at 1,000 kg (20 cwt) and the finisher used to complete the feeding programme. For birds killed at larger weights, that is, capons and roasters, the starter is fed for 750 kg (15 cwt) per 1,000 birds, the grower for 1,000 kg (20 cwt) per 1,000 birds, the finisher for 1,000 kg (20 cwt) per 1,000 birds, followed by the finisher No. 2. For heavyweight roasters and capons the age at killing is usually between ten and fourteen weeks, depending upon the weight of the bird required. The protein, amino acid and vitamin specifications for the finisher No. 2 are 80% of the finisher No. 1 levels. Male and female broilers are often grown separately. Because of the slower growth rate of the female its requirements for nutrients is less than the male, and although specific female rations are feasible, it is more usual to make amendments to hatched feeding programmes to cater for the slower growing female. Smaller quantities of the starter and grower rations being fed with levels of around 95% of the hatched starter allowance and 85% of the hatched grower ration allowance. For example, the female programme would be 950 kg (19 cwt) instead of 1,000 kg (20 cwt) of starter per 1,000 birds and the grower 1,275 kg (25½ cwt) instead of 1,500 kg (30 cwt) per 1,000 birds. When the cost of protein is very high, separate sex feeding is at its most economic.

Turkey rations

Relative to bodyweight gain the nutrient requirement of broilers and turkeys is very similar. The turkey is fed high levels of protein and amino acid during the early period of growth compared to the broiler. This is necessary in order to maintain a fast growth rate and, coupled with high levels of vitamins and trace elements, to obtain low mortality and freedom from leg disorders. Turkeys are marketed at a variety of ages depending upon the specific market requirement. The relatively small 4–6 kg ($8\frac{3}{4}$–$13\frac{1}{2}$ lb) oven-ready turkey is a bird marketed at around ten to twelve weeks of age, whilst the more traditional Christmas weight bird, between 7–8 kg ($15\frac{1}{2}$–$17\frac{3}{4}$ lb), is marketed between sixteen and twenty weeks of age. Older birds for the catering trade may vary between 10–13 kg (22–$28\frac{3}{4}$ lb). Turkeys are efficient converters of food into meat, and in order to capitalize on this ability, particularly in those birds marketed at an early age, the nutrient levels of the rations should be maintained at a high level. A typical programme for birds killed between ten and twelve weeks of age involves the turkey starter for the first four to five weeks, followed by the rearer fed to between eight and nine weeks, and a high efficiency early finisher ration fed to twelve weeks of age. Where the birds are marketed at ten weeks of age the second stage ration is discontinued at eight weeks. Turkeys marketed at more mature ages can be fed less concentrated rations in the early period as any slowness in growth rate in this early period can be compensated for during the later period of growth. This phenomenon, known as compensatory growth, means that less expensive starter and second stage rations need be fed or, probably more practical, more expensive food can be fed for a shorter period of time. Turkeys marketed at an early age tend to lack so-called bloom or finish, which is the deposition of a layer of subcutaneous fat. This type of finish is desirable as it imparts a creamy texture to the skin. Such birds are naturally immature, and in order to encourage this 'finish', the early turkey finisher ration fed for the final period of growing normally contains both high protein and high energy. Added fat is often used to achieve the energy which is necessary.

Duck rations

Weight for age, ducks grow faster than any other species or class of poultry. In forty-nine days a duckling may reach an average weight of 3–$3\frac{1}{2}$ kg ($6\frac{1}{2}$–$7\frac{1}{2}$ lb). Whilst their ability to convert food into meat is good, it is not as efficient as the broiler. Unlike either chicken or turkey the duck has the natural ability to lay down levels of subcutaneous fat which give it the desirable bloom or finish. With birds marketed at forty-nine days of age, the starter is fed between fourteen and twenty-one days, followed by the grower to thirty-five days, followed by the finisher ration. Two stage programmes are also satisfactorily used with the starter being fed for the first twenty-one days. Ducklings show little response to the commonly used chemical growth promoters and these are not used in duck rations. Fast growing ducklings are particularly susceptible to leg weakness and in order to prevent this, high levels of B group vitamins, particularly nicotinic acid, are necessary. As with all fast growing birds, the calcium/phosphorus ration and level of vitamin D_3 is critical.

Guinea-fowl rations

Although guinea-fowl can satisfactorily be grown on a feeding programme involving a chick starter followed by a grower ration, much more efficient and economic growth is obtained through the feeding of rations having similar nutrient specifications to those fed to broilers. The smaller or domestic poultry-keeper is able to produce this bird to market weight with rations of lower nutrient density. Guinea-fowl should not be fed rations containing the coccidiostat Monensin, as it has been found that at levels which are used for the control of coccidiosis in broilers the coccidiostat is toxic for this species.

Food consumption guides

The following tables have been prepared not as an absolute indication of how much food poultry will eat, but to serve as a guide to the quantities which are eaten under average conditions. Since no two farms have the same conditions, the average is a misnomer and there will almost certainly be a degree of difference between the figures presented and those from a specific unit.

Feedingstuffs analysis data

Material	Prot.	Oil	Fibre	Lys.	Meth.	M & C	Tryp	Threo	ME Cals/kg	Ca.	Phos	Av. P	Salt
BARLEY	10.0	1.5	4.5	0.35	0.15	0.36	0.12	0.35	2750	0.10	0.40	0.12	0.10
MAIZE	9.0	3.8	2.7	0.21	0.18	0.34	0.09	0.39	3417	0.02	0.29	0.10	0.10
WHEAT	11.0	1.9	2.6	0.30	0.14	0.36	0.12	0.31	3090	0.05	0.30	0.10	0.10
OATS	11.0	4.5	11.0	0.40	0.18	0.38	0.15	0.40	2513	0.10	0.35	0.12	0.10
WHEATFEED	15.50	3.80	7.0	0.60	0.18	0.38	0.16	0.44	1984	0.10	0.80	0.24	0.05
FISH '66'	66.0	4.0	1.0	4.60	1.60	2.40	0.66	2.60	2750	7.5	3.5	3.5	1.5
HERRING	70.0	9.0	1.0	5.40	2.00	2.70	0.70	2.75	3200	3.0	2.0	2.0	1.5
MEAT 50/10	50.0	10.0	2.5	2.40	0.68	1.28	0.28	1.60	2560	8.0	4.0	4.0	1.5
POULTRY OFFAL MEAL	60.0	22.0		2.50	1.00	1.70	0.53	2.00	3630	1.50	1.00	1.00	0.70
SKIM MILK	33.0	0.90		2.30	0.86	1.20	0.45	1.51	2510	1.25	1.00	1.00	1.40
DPM (Battery)	25.0	2.0	15.0	0.32	0.11	0.32	0.53	0.40	770	7.4	2.0		1.00
MAIZE GLUTEN '26'	26.0	2.5	8.0	0.70	0.35	0.70	0.16	0.80	1765	0.05	0.80	0.27	0.20
MAIZE GLUTEN '60'	61.0	2.5	1.5	0.90	1.60	2.60	0.30	2.00	3660	0.01	0.40	0.12	0.05
SOYA EXT	44.0	1.0	7.0	2.90	0.65	1.30	0.66	1.80	2240	0.25	0.60	0.24	0.05
SOYA (Full Fat)	36.2	18.0	4.4	2.29	0.55	1.09	0.54	1.45	3307	0.17	0.47	0.17	0.07
GROUNDNUT Ext.	52.0	1.3	3.9	1.73	0.44	1.16	0.49	1.31	2535	0.20	0.60	0.18	0.25
FIELD PEAS	23.0	2.0	6.0	1.56	0.25	0.60	0.21	0.94	2425	0.06	0.40	0.13	0.07
FIELD BEANS	26.0	1.5	7.5	1.65	0.20	0.62	0.23	0.94	2425	0.10	0.40	0.13	0.05
GRASS MEAL	15.5	2.8	26.0	0.80	0.35	0.60	0.20	0.25	1150	0.90	0.30	0.12	0.50
LIMESTONE										38.0			
DICALCIUM PHOSPHATE										26.0	18.0	18.0	
TALLOW		100.0							7275				

Recommended micro-nutrient levels for poultry

	Broiler Starter	Broiler Grower	Broiler Finisher	Chick Starter	Grower 1	Grower 2	Layer HND	Hybrid Breeder	Broiler Breeder 1	Turkey Starter
Micronutrients										
VITAMIN A mIU	12.0	12.0	12.0	10.0	8.0	8.0	6.0	10.0	12.0	16.0
VITAMIN D3 mIU	4.0	4.0	4.0	3.0	2.4	2.4	3.0	3.0	3.0	5.0
VITAMIN E mg/kg	10.0	10.0	5.0	6.0	4.8	4.8	4.0	8.0	10.0	12.0
VITAMIN K mg/kg	2.0	2.0	2.0	2.0	1.6	1.6	2.0	2.0	2.0	4.0
RIBOFLAVIN (B2) mg/kg	5.0	5.0	5.0	5.0	4.0	4.0	3.0	6.0	8.0	10.0
THIAMINE (B1) mg/kg	1.0	1.0	1.0	0.5	0.4	0.4	0.5	1.0	1.0	1.0
PYRIDOXINE (B6) mg/kg	1.0	1.0	1.0	0.5	0.4	0.4	1.0	3.0	4.0	2.0
COBALAMIN (B12) mcg/kg	10.0	10.0	10.0	6.0	4.8	4.8	3.0	6.0	6.0	8.0
BIOTIN mcg/kg	80.0	80.0	—	50.0*	50.0*	50.0*	—	50.0	50.0	80.0
FOLIC ACID mg/kg	1.0	1.0	0.5	0.5	0.4	0.4	—	1.0	2.0	2.0
CHOLINE mg/kg	200.0	200.0	200.0	200.0	160.0	160.0	—	150.0	150.0	35.00
NICOTINIC ACID mg/kg	30.0	30.0	20.0	20.0	16.0	16.0	8.0	25.0	25.0	40.0
PANTOTHENIC ACID mg/kg	8.0	8.0	6.0	8.0	6.4	6.4	3.0	8.0	10.0	12.0
MANGANESE mg/kg	70.0	70.0	100.0	65.0	52.0	52.0	70.0	70.0	70.0	80.0
ZINC mg/kg	50.0	50.0	60.0	40.0	32.0	32.0	40.0	50.0	60.0	60.0
IRON mg/kg	25.0	25.0	20.0	15.0	12.0	12.0	20.0	20.0	20.0	30.0
COPPER mg/kg	10.0	10.0	8.0	8.0	6.4	6.4	6.0	8.0	8.0	10.0
COBALT mg/kg	0.5	0.5	0.5	0.5	0.4	0.4	0.5	0.5	0.5	0.5
IODINE mg/kg	1.0	1.0	1.0	1.0	0.8	0.8	1.0	1.0	1.0	1.0
SELENIUM mg/kg	0.1	0.1	0.1	0.1	0.08	0.08	0.1	0.1	0.1	0.1
MOLYBDENUM mg/kg	0.5	0.5	0.5	—	—	—	—	0.5	0.5	

* Biotin recommended for brown birds

Recommended nutrient allowances for poultry

		Broiler Starter	Broiler Grower	Broiler Finisher	Capon	Chick Starter	Grower 1	Grower 2	Layer HND	Layer MND	Layer LND	Hybrid Breeder
PROTEIN	%	23.0	20.0	19.0	17.0	18.0	15.0	13.5	18.0	17.0	16.0	17.0
ME kcals/kg		3056	3078	3100	3056	2825	2722	2722	2834	2786	2746	2786
ME MJ/kg		12.79	12.88	12.97	12.79	11.82	11.39	11.39	11.86	11.66	11.49	11.66
TOTAL LYSINE		1.30	1.10	1.00	0.75	0.86	0.60	0.56	0.82	0.78	0.74	0.78
AV. LYSINE		1.18	1.00	0.90	0.68	0.78	0.54	0.50	0.74	0.70	0.67	0.70
METHIONINE		0.50	0.47	0.45	0.44	0.40	0.30	0.24	0.38	0.36	0.34	0.36
M+C		0.86	0.82	0.78	0.70	0.65	0.50	0.45	0.64	0.60	0.56	0.60
CALCIUM		0.90	0.90	0.90	1.00	0.90	0.90	1.00	3.75	3.55	3.45	3.45
AV. PHOSPHORUS		0.45	0.45	0.43	0.43	0.40	0.30	0.33	0.33	0.33	0.33	0.33
TOTAL PHOSPHORUS		0.70	0.70	0.68	0.68	0.65	0.55	0.50	0.56	0.60	0.62	0.65
TOTAL SALT		0.38	0.40	0.40	0.40	0.38	0.38	0.38	0.38	0.40	0.40	0.40
SODIUM		0.15	0.15	0.15	0.15	0.15	0.15	0.15	0.15	0.13	0.13	0.13

Turkey Rearer	Turkey Grower	Early Finisher	Finisher	Turkey Breeder	Duck Starter	Duck Grower	Duck Finisher	Duck Breeder
12.0	9.6	12.0	9.6	16.0	12.0	10.0	10.0	12.0
5.0	4.0	5.0	4.0	3.0	4.0	4.0	4.0	3.0
10.0	8.0	10.0	8.0	20.0	30.0	20.0	20.0	30.0
3.5	2.8	3.5	2.8	2.0	2.0	2.0	2.0	2.0
8.0	6.4	8.0	6.4	12.0	8.0	6.0	6.0	8.0
0.5	0.4	0.5	0.4	1.0	1.0	1.0	1.0	1.0
0.5	0.4	0.5	0.4	2.0	1.0	1.0	1.0	1.0
6.0	4.8	6.0	4.8	8.0	6.0	6.0	6.0	8.0
50.0	40.0	50.0	40.0	50.0	80.0	80.0	80.0	80.0
1.5	1.2	1.5	1.2	3.0	1.0	0.5	0.5	1.0
200.0	160.0	200.0	160.0	200.0	200.0	200.0	200.0	300.0
35.0	28.0	35.0	28.0	50.0	50.0	40.0	40.0	40.0
10.0	8.0	10.0	8.0	15.0	15.0	12.0	12.0	15.0
70.0	56.0	70.0	56.0	80.0	70.0	70.0	70.0	80.0
60.0	48.0	60.0	48.0	70.0	60.0	60.0	60.0	60.0
25.0	20.0	25.0	20.0	20.0	20.0	20.0	20.0	20.0
8.0	6.4	8.0	6.4	8.0	10.0	8.0	10.0	10.0
0.5	0.4	0.5	0.4	1.0	0.5	0.5	0.5	0.5
1.0	0.8	1.0	0.8	0.5	1.0	1.0	1.0	1.0
0.1	0.08	0.1	0.08	0.1	0.1	0.1	0.1	0.1
—	—	—	—	0.5	—	—	—	0.5

Broiler Breeder 1	Broiler Breeder 2	Turker Starter	Turkey Rearer	Turkey Grower	Early Finisher	Finisher	Turkey Pre-breeder	Turkey Breeder	Duck Starter	Duck Grower	Duck Finisher	Duck Breeder
16.5	15.5	28.0	23.0	18.0	20.0	14.0	14.0	18.0	20.0	17.0	15.0	17.5
2722	2710	2987	2987	2947	3140	2947	2698	2834	2987	2987	2987	2722
11.39	11.34	12.50	12.50	12.33	13.14	12.33	11.29	11.86	12.50	12.50	12.50	11.39
0.76	0.72	1.80	1.30	0.90	1.00	0.62	0.53	0.80	1.15	0.90	0.72	0.80
0.68	0.65	1.62	1.20	0.80	0.90	0.56	0.48	0.73	1.04	0.82	0.85	0.72
0.35	0.33	0.80	0.60	0.35	0.60	0.30	0.28	0.38	0.60	0.52	0.42	0.36
0.58	0.54	1.10	0.85	0.68	0.85	0.50	0.46	0.64	0.88	0.74	0.63	0.60
3.00	2.50	1.20	1.00	1.00	0.90	1.00	1.00	2.80	1.00	1.00	1.00	2.50
0.35	0.35	0.65	0.55	0.45	0.45	0.40	0.33	0.40	0.55	0.55	0.55	0.50
0.65	0.65	0.80	0.80	0.75	0.75	0.70	0.65	0.70	0.75	0.75	0.70	0.65
0.40	0.40	0.32	0.32	0.32	0.32	0.32	0.40	0.40	0.32	0.32	0.32	0.38
0.13	0.13	0.13	0.13	0.13	0.13	0.13	0.13	0.13	0.13	0.13	0.13	0.13

11 Health

There is just one thing worse than a dead bird, and that is a sick one. It is a misery to itself, a worry to its owner and, from a commercial viewpoint, a 'passenger' consuming space, food and time for little or no return. Unfortunately, the financial value of an ailing bird, in most cases, does not justify calling in a vet. The commercial producer will probably cull it dispassionately while the small flock owner is apt to spend a disproportionate amount of money in trying to cure it. Yet most health problems can be put down to ignorance or neglect in the fields of hygiene, environment, nutrition or vaccination.

It is very easy, if you are a domestic poultry-keeper, to believe that because the birds are kept in 'natural' conditions, with access to sunlight and an exercise area, they are somehow hardened to resist infections. Few things could be further from the truth. The sun's ultra-violet rays are beneficial, but alone they are no substitute for hygienic conditions, and the majority of setbacks in small flocks can be traced directly to mud and poor management.

The first thing to question is whether fresh stock are being put into surroundings recently occupied by other birds. An earlier flock may have shown no clinical signs of disease before it was removed, but younger birds, abruptly exposed to the pathogens and parasites left behind by their predecessors, have no chance to adjust to the challenge.

Internal parasites

Coccidia and worms are the front-line enemies on free range or deep litter. These are the parasitic infections easily picked up by foraging poultry of all species, including the hardy guinea-fowl.

In general the life cycle of worms can be broken by resting the ground for a period between flocks—which is why alternating between two runs is a good idea—or by totally digging out the top 5 cm (2 in) of clinker and ash in a made-up run and replenishing it. Litter in deep litter houses may be fumigated and topped up with additional material, but it is safer to replace it completely.

Several types of worm can attack poultry, and since they only occasionally cause deaths, their presence may be overlooked, although it would take a very unobservant, or very inexperienced poultryman not to notice that something was wrong. All intestinal worms, if allowed to build up in numbers, cause loss of condition, slow or non–existent weight gain and, in layers, poor productive performance.

Roundworms
The large roundworm is white, inhabits the small intestine and may be up to 10 cm (4 in) long. It can therefore be fairly easily identified in the droppings. Piperazine compounds, dosed at the rate of 300–500 mg/kg of bodyweight, or added to the drinking water at 2–4 g/gal have been found to be effective.

Caecal worms
A small worm, up to 12.5 mm ($\frac{1}{2}$ in) in length, can be found in the caeca, or blind gut, of birds. Unless it reaches a quantity which can cause inflammation of the caeca it is not too troublesome in fowls, but it can have an important economic effect in turkeys, because it in turn harbours the protozoan *Histomonas meleagridis*. This protozoan, which can live in both the worm and its eggs, leads to 'blackhead', a killer disease in turkeys, which particularly endangers growing birds at around six to ten weeks. Yellow droppings and a hunched, depressed appearance are characteristic symptoms, but despite the name, darkening of the head is not always present. These days it is largely controlled in commercial flocks by the addition of efficient drugs as a premix in feeds. Several treatments, available only under veterinary prescription in the UK, can be administered through the feed or drinking water, and tablets can be obtained to dose individual birds. But the disease is best avoided by keeping the birds off potentially infected surfaces, and not running turkeys with fowls.

Threadworms
Minute worms which live in the upper part of the intestine can cause trouble in the form of debilitation and diarrhoea if infestation reaches a chronic level. Again a feed premix or soluble powder is the treatment.

Gizzard worm
A threadlike worm from 12–25 mm ($\frac{1}{2}$–1 in) makes its home in the unpromising surroundings of the gizzard in geese. Growing birds are most prone to suffer from it and show symptoms in a slow, staggering gait, weakness and lack of weight. Treatment is a 1-ml capsule of carbon tetrachloride for goslings up to 3.6 kg (8 lb) and a 2-ml capsule for bigger birds, repeated after seven days. Again, infestation is more likely where birds are run over heavily-used ground, perhaps with only a small communal pond.

Tapeworms
Probably the best known bowel parasite is the tapeworm, a flat, seg-

mented creature whose life-style is a favourite subject among biology teachers. Several species can exist in poultry, the largest reaching up to 25 cm (10 in) in length.

Each segment of a tapeworm is virtually a developing egg case which breaks off and passes out of the host when it is ripe. To continue the existence, the eggs from the segment must be taken in by earthworms, slugs, snails, beetles or flies. If one of these carriers is then eaten by poultry, a new infestation is set up in the gut. These parasites are difficult to eliminate once they have established a hold in the gut wall.

Gape worms

Gasping, coughing and signs of partial suffocation in young birds can be a sign of an infestation of gape worms. These bright red worms live in the trachea or wind-pipe and have a characteristic double-headed appearance, the smaller male worm being attached to its female counterpart, which may be 15–25 mm ($\frac{5}{8}$–1 in) long.

Eggs from the gape worm are picked up from the ground direct, or may be ingested by slugs or earthworms subsequently eaten by birds. Symptoms can take a week or so to show up after the initial infection.

An old short-term remedy for gape worms, still appropriate for individual birds, is to take a feather, dip it in turpentine, insert it in the wind-pipe and twist it. On withdrawal the bird will cough up some of the worms.

Modern drugs are certainly more effective than some of the singularly messy treatments once recommended, but some parasites—notably the tapeworm—once established can be very difficult to dislodge. Observing a few elementary rules will greatly diminish the risk of their gaining a hold.

> Never run fresh birds on stale ground or litter just vacated by older stock.
> Clean and disinfect all housing and equipment between flocks.
> Keep stock out of reach of potential carriers—wild birds, beetles, earthworms, slugs and snails.
> Monitor the background of new birds, ensuring that any grown stock brought in come from clean surroundings.

There is no excuse for ignoring these rules where poultry are enclosed or allowed a small run. With free range the difficulties are admittedly greater, but any serious outbreak of worms is still an indication that somewhere along the line the poultry-keeper is failing, in hygiene techniques, stock observation, or both.

Coccidiosis

Among the more widespread and economically significant poultry diseases, coccidiosis probably ranks highest because of its considerable impact on the broiler industry. It mainly affects growing birds, and survivors will always be carriers. Since, like worms, it is a parasitic complaint spread in the droppings, growers and broilers raised on the floor, rather than in cages, are at greatest risk.

Coccidia are microscopic protozoa which attack a wide range of animals. Early signs that they are at work in a young poultry flock may show in birds becoming hunched and mopey. Feathers may be ruffled, and blood-streaked diarrhoea is common. These symptoms are not conclusive on their own. Diarrhoea, for example, with pasted vents and soiled feathers, can result from a number of disorders, such as chilling or incorrect feeding, so before jumping to a hasty conclusion, make sure the birds are not being confined in damp or draughty conditions or suffering nutritionally. The disease is sufficiently common to be quickly suspected, however, especially if deaths start to occur. Mortality, in a susceptible flock receiving no medication, can easily run as high as 75%.

As a flock ages, under natural conditions, it tends to develop resistance to mild challenges and in older birds, if symptoms show at all, they tend to be simply diarrhoea, often without signs of blood. This is not to say that birds, of whatever age, will withstand the parasite when exposed to it for the first time, and it will cause deaths. It is rarely found in caged layers.

Nearly all the coccidia which affect poultry are of the genus *Eimeria*, which is abbreviated to the prefix *E.* by scientists. Thus, chickens can be affected by *E. tenella*, *E. necatrix*, *E. brunetti*, *E. maxima* or *E. acervulina*, and each of these species favours a different part of the gut, or shows different lesion characteristics in the intestinal wall. *E. tenella* concentrates in the caeca, where it causes whitish patches and pin-point haemorrhages for example, while *E. necatrix* causes a lot of damage in the mid-part of the small intestine, with extensive haemorrhages, blood and mucus in bad cases. *E. acervulina*, which appears mostly in the duodenum, has variations which sometimes make it difficult to distinguish from *E. mivati* and *E. mitis*, two additional forms, and these three tend to be regarded as one group. One further species, *E. praecox* is recognized in chickens, but is not lethal, as the others can be.

E. meleagrimitis and *E. adenoeides* are the commonest forms of coccidia found in turkeys in Britain, the first attacking mainly the duodenum and jejunum, the second the caeca and lower intestine. But this bird can also be affected by *E. meleagridis*, which has a passing effect and is not a cause of serious disease, and *E. gallopavonis*, which also causes few losses and does not appear to occur in Europe. Turkeys in general, after about the first four weeks of life, seem to resist coccidiosis better than chickens.

Guinea-fowl, especially when housed like broilers, are prone to *E. numidae* when young and are also subject to *E. grenieri*, which is generally less damaging.

For ducks, an exception to the *Eimeria* genus, *Tyzzeria perniciosa* is the most pathogenic species in the UK, followed by *E. anatis* and *E. danailovi*. The strongest challenge to geese is *E. truncata*, unusual in that it occupies the kidneys. Mortality can be high. Another killer is *E. anseris* while goslings particularly may also suffer from a disease caused by *E. nocens*.

This long list of Latin names illustrates the very broad disease spectrum veterinarians have in mind when they speak of 'coccidiosis'. In the laboratory the parasites are categorized by removing the intestines, studying the damage, opening them and making microscopic tests. On the outcome of these, they can decide the most appropriate coccidiostat to include in feeds to keep the parasites in check.

There is nothing to stop the small-scale practical poultryman from conducting post mortems. Examination of the gut will allow at least a crude diagnosis, based on the symptoms described here, and sufficient experience can be gained from opening dead birds to build a broad impression of the more obvious avian ailments. But the final arbiter will always be the laboratory investigator, of course, who has the knowledge and facilities at his fingertips.

Despite their variations, all coccidia share the same basic pattern of existence. Being of microscopic size, they invade the cells, usually of the gut lining, destroying them. When, by multiplication, the coccidia have grown and developed into male and female forms, they unite to produce eggs or oöcysts which are excreted in the bird's droppings in their millions. These minute oöcysts have a tough outer casing and are highly resiliant, surviving for months on the ground if need be, within a temperature latitude of roughly 10–50°C (50–122°F). They enjoy damp conditions and will stand up to many disinfectants, but ammonia and methyl bromide are lethal to them. Oöcysts take about two days on the ground to sporulate and become infective to a new host. Once eaten, the case of the oöcyst is dissolved by digestive juices, releasing several parasites to attack, multiply and repeat the cycle.

Prevention is really a question of keeping birds away from the source of infection, namely droppings and contaminated ground. Where young stock are run on the ground or on a house floor, assume that oöcysts are present even if no symptoms are showing and, to give them no chance to build up, move the growers to fresh ground, or change litter between every flock.

Under intensive conditions, with many thousands of birds, however, extra precautions must be taken and special anticoccidial drugs have been developed which are metered into compound rations in tiny quantities, sometimes no more than three parts per million.

Different drugs act in different ways within the life cycle of the parasite to prevent its development. The trouble is that coccidia tend to become resistant to specific drugs over a period, and flocks must be continuously watched by veterinarians to determine whether an anticoccidial is still doing its job, and a wide range of anticoccidials has therefore been developed.

Among the drugs now available, Elancoban (known as Monensin in the USA) has captured a large part of the market as an additive for broiler feeds. Others include May & Baker's Embazin for all types of poultry, Dow Chemical's Lerbek, Merck, Sharp & Dohme's Pancoxin and Cyanamid's Cycostat 66. There are many others.

Cleaning

Whatever the size of the poultry unit, hygiene is the spearhead to health. On the small unit a methodical approach can be quick, cheap and elementary. But it *must* be done, and too often back-garden flocks are the ones to suffer most from dirty conditions and pests such as mites and lice, due to the negligence of their owners.

As soon as a flock is removed from a house all the portable equipment should be brought out for cleaning. Professionals will use a high pressure washer or steam hose, but even the amateur need not stint on boiling water and some washing soda. Protect your hands with strong rubber gloves.

Scrub the perches—which should be detachable—immerse the feeders and drinkers in hot water and clean out the nest boxes, destroying the old bedding materials. Droppings boards should also come out so that they can be scrubbed down.

If the same run has to be used for successive flocks, skim off the existing surface, lime and harrow the subsoil and resurface with ash and clinker.

Disinfection is the next step; a wise precaution even on the backyard unit. It adds only slightly to costs and is likely to repay these many times over in the health of the new flock.

Today's choice is wide, ranging from detergent sterilizers which both clean and sanitize, to straightforward disinfectants used after a washdown. Some can be used in sprays while birds are in the house, others need a period of up to seventy-two hours before stock are housed. Certain compounds are primarily for use on floors and litter, others have a more general application. Nearly all are used in diluted form, but the diluent might be water, kerosene or gas oil.

Choice is a matter of deciding precisely what is required of the disinfectant in relation to the size and nature of the unit. One fundamental question, for instance, is whether the compound can be bought in convenient quantities. If just a small proportion is used, can the rest be stored over any period without losing its strength? Storage is an important consideration even on large units, of course. Powders are more compact than liquids.

For earth of litter floors, disinfectant sprays diluted with oil or kerosene, which inhibit bacteria, moulds and coccidia, are frequently used. Building interiors, utensils and footwear can be treated with a general disinfectant. Choose one with a broad 'spectrum', having killing strength against bacteria, viruses, fungi and moulds, even in the presence of organic matter. Disinfectants must never leave residues toxic to livestock, of course, although some may need the operator to wear protective clothing and a mask when applying them. Ideally they should be non-irritant, non-staining and if they also act as insect repellants, so much the better. Where ingrained dirt is not present the non-toxic chlorine, iodine and quaternary ammonium type compounds are popular.

Detergent disinfectants save time, and cleaning gangs frequently use them in pressure washers, but they are limited to cleaning roles. They are not suitable as fogging or fumigation agents.

Occasionally, as when outbreaks of disease have occurred, an intermediate disinfection is recommended while a flock is still in the house, sprayed into the air and around the birds. Certain disinfectants can be used in this way, and some, in highly diluted form, can be used to sterilize food and water supplies for bird consumption, but the golden rule in every case is *read the instructions thoroughly*. More mistakes with disinfectants arise from hasty reading, or even non-reading, of instructions than any other cause. Observation of dilution rates, storage conditions and toxic levels is all-important. Apart from avoiding possibly unpleasant side-effects, correct usage saves throwing money, literally, down the drain by ineffective sanitation.

Many disinfectants can be bought straight off the shelf, of course, but fortunately the purchaser does not have to rely purely on the pack label for guidance. Lists of approved disinfectants are drawn up by Government departments. In England this goes under the rather complicated title of *Schedule to the Diseases of Animals* (*Approved Disinfectants*) (*Amendment*) *No. 2 Order 1976* obtainable from Her Majesty's Stationery Offices.

General purpose disinfectants are listed in parts 4 and 5 of the Schedule while disinfectants approved specifically for use against Newcastle disease (fowl pest) are listed in part 2. With this type of disease an Order or Regulation may be made under the Diseases of Animals Act 1950 requiring the disinfection of the premises with an 'approved' disinfectant. Where products have received approval, makers will generally state this on the pack.

In the US both the Food and Drug Administration and the USDA co-operate to 'clear' a drug or chemical for use on poultry or in poultry houses. Producers in America must check this list frequently as it is constantly changing. Officials of these agencies regularly check samples of eggs and poultry.

On large commercial sites there is something to be said for contract cleaning. Specialist firms have the right equipment, tractors, pressure washers, vacuum cleaners, knapsack sprayers and the like, with a trained team to use it. They are experienced in handling the disinfectants and know what precautions to take and the type of protective clothing to wear. Nevertheless, some poultry companies have enough sites to justify their own equipment, or prefer to lease it just when required, and employ their own labour.

Time is money on a busy farm, and empty poultry houses are not earning, so there is pressure to re-stock them as quickly as possible. Most professionals know the importance of not keeping different aged birds on the same site, but if a producer decides to run on a flock of layers for a second season, for instance, an overlap may be unavoidable and the all-in all-out code will be broken. Hygiene programmes must stand up to this sort of challenge effectively. With detailed variations according to site, the commercial cleandown follows a similar logical pattern to that of the domestic poultry-keeper, starting with the removal of equipment from the houses. Then the pattern is broadly as follows:

> Remove litter and manure. In cage houses, especially deep pit systems, manure may be allowed to lie if it is well isolated from the birds, but you take the chance of flying insects bringing infection to a new flock unless the manure is sprayed.
>
> Brush down cobwebs and remove dust from roof assemblies, lights and anywhere else it may have collected.
>
> Remove inlet baffles if possible, brushing and/or vacuuming dust away.
>
> Take out fans, cleaning them and their surrounds.
>
> Remove surplus feed from bulk bins.
>
> Bring in pressure washing equipment. Even cold water at over 40 kg/cm² (500 lb/in²) will often shift caked-on dirt, but higher pressures and/or hot water and detergents are available with most modern equipment to deal with organic debris. Make sure that wall and ceiling linings can take the pressure—*and that all electrical fittings are avoided and properly neutralized.* A number of deaths have occurred through mistakes here.
>
> Put a sterilant, at the recommended level, into the header tanks of the drinking system. Allow it time to act, then drain down through the drinker lines, to sterilize the whole system.
>
> Using a knapsack or aerosol-type generator, spray the house, feed store, and standing equipment with an approved disinfectant.
>
> Re-install fans.
>
> Replace the movable equipment, which should meanwhile have been steam-cleaned or pressure washed and, if possible, totally immersed in a disinfectant solution.
>
> Some poultry-keepers, particularly rearers and broiler growers, take the additional precaution of fumigating the house. Seal all openings, use the heaters to raise the temperature and spray the atmosphere with water to raise humidity, then proceed as page 156.

A hygiene programme does not cease with cleaning down, however. Between cleaning sessions diseases, which travel very happily on foot or by lorry as well as by air, must be kept away from the birds. Practically all large-scale poultry farms now have a vehicle dip, containing disinfectant, at the entrance, and a footbath of a similar compound at the door of each house. Unfortunately, these are frequently allowed to degenerate into sludge-filled troughs which are worse than useless, because they give a false sense of security. Replenish disinfectant dips regularly. Keep visitors to a minimum and provide boots or plastic overshoes for those who do come. Disposable coats can also be supplied. If it is necessary to visit flocks of different ages, even farm personnel should, whenever possible, go to the younger birds first.

External parasites

Undoubtedly the most directly troublesome pests for the birds are external parasites, mites, lice and fleas. These may contribute to the transmission of certain diseases between stock, but their main effect on health is in the sheer irritant misery they bring, damaging skin and plumage, disturbing sleep and debilitating the bird.

Always take any sign of external parasites seriously. A few today could be tens of thousands, perhaps millions, in about a fortnight, so prompt action is called for. Small flock owners have the opportunity to examine all new birds on arrival and this should be done thoroughly. Look for ragged feathering or areas of missing feathers. Inspect the vent area and part the feathers on the body. Watch for moving parasites but look also for scabs on the skin, white powdery deposits or small grey clumps— which are actually eggs—attached to the base of feathers.

Infested birds must be isolated and treated before they are introduced to clean quarters. In fact it is wise to dust even apparently clean birds with insecticide. By the same token, of course, the newcomers should not be exposed to infested surroundings. Given a few cracks and crevices, parasites that survive a careless house-cleaning will be there to greet their new hosts.

Individual bird inspection is impracticable for large flocks, but samples can be checked, and modern spraying devices make it possible to deal with poultry collectively, as described below. Lice, mites and fleas all tend to be susceptible to the same chemical agents.

Lice
These are insects, usually about 3 mm ($\frac{1}{8}$ in) long, ranging in colour from off-white to dark grey and brown, which vary slightly in type and habits. Wing lice and shaft lice eat mainly fragments of feathers and live generally among the plumage. Body lice depend more on particles of skin and live on the body surface, scarifying it and causing irritation and scabbing.

Mites
More versatile, and therefore more persistent than lice, mites are spider-like creatures with eight legs. They suck the blood of their host, create a lot of irritation, and in extreme concentrations can cause anaemia and death. Again several varieties exist, the two with the greatest effect on poultry being the red mite or roost mite, *Dermanyssus gallinae*, and the northern fowl mite, *Ornithonyssus sylvarum*.

In checking for red mite infestations, the house as well as the bird needs to be inspected, because this species spends the day massed in any convenient dark corner, split insulation linings or cracked woodwork being ideal, and invades the birds at night. It can survive for several months in a house, even when birds have been removed.

Northern fowl mites spend most of their time on their host, although they will last as much as a month away from the birds. It is not uncommon to see them first as moving dots on a hen's eggs or on cage frameworks. Always check the males in a breeding flock as well as the females. Under certain conditions of temperature, and when dark females are mated to light coloured males you may see a heavy infestation of Northern fowl mite on the males and very few on the females. These mites, on the male, can have a damaging effect on fertility as they tend to congregate in the area of the vent and produce a scabby condition which discourages the male from mating.

Another mite, *Knemidocoptes mutans*, causes a condition known as 'scaly leg' in which the scales become rough and deformed by the presence of the parasite between them. This thickening and encrustation of the legs and feet can make it difficult for the bird to perch or walk, but it can be effectively treated by dipping the affected parts in kerosene and linseed oil, mixed in the proportion $1:2$.

Fleas
Although fleas are quite a separate genus from mites, some of their habits are similar. They too are blood-suckers who cause itching and restlessness to their victims, and like the red mite they lay their eggs in any suitable area, such as nest box litter, and have a survival period of months in empty houses if necessary.

Ticks
Yet another vampire is the tick, *Argus persicus*, which hides in the house in daytime and raids the birds at night, although the larval stages may remain attached to their host for long periods. Usually these creatures simply cause anaemia and a possible loss of production in poultry, but a report in the American magazine *Poultry Digest* of July 1975 noted a case of paralysis among twenty week-old chickens which the Pennsylvania Bureau of Animal Industry put down to a tick infestation.

Total prevention of pests of this kind is almost impossible. Sooner or later, via contaminated vehicles or crates, contact with wild birds or something else beyond the poultryman's control, an infestation will start. Much can be achieved by being alert to them, though. Red mite in the roosting areas of small houses can be prevented from reaching their hosts, for example, by painting the perches with nicotine sulphate, or dipping the perch ends in kerosene each week. Dust the perch sockets with a reliable insecticide.

Where pests are found after birds have been removed from a house, cleaning, as described before, is the first step, followed by treatment with a pesticide. A number of preparations are on the market, many containing malathion, which can be used whether or not birds are present. Sevin, the active ingredient of which is a chemical called carbaryl, also appears frequently on pesticide labels. The name Sevin is a registered trade mark of the Union Carbide Corporation, USA. One product, Coumaphos or Co-Ral, supplied by Bayer Products, is widely used for dusting or spraying against lice and mites in the US but has no product licence for use with poultry in the UK, although it is allowed in sheep dips. Other compounds active against poultry pests include certain organo-chlorine substances but these are not universally approved and should only be used where other pesticides are unsuitable. Do not apply them to litter destined to be spread on pasture-land or added to cattle feeds.

In empty houses, pesticides may be applied, as per maker's instructions, in spray, powder, smoke or vapour form. Some methods require special applicators such as the system of cartridges. The efficiency of any method depends on the circumstances; the type—and often the age—of house, the degree of infestation and the nature of the pest. If a lot of awkward corners are involved, fumigation might be resorted to, but clean down thoroughly beforehand. Fumigants do not readily penetrate debris. Sulphur dioxide, which can be produced by burning sulphur, is sometimes used as a method of fumigating small poultry houses. The interior of the building needs to be wetted for this to work. Professional cleaning gangs employ methyl bromide, a highly dangerous fumigant in inexperienced hands, but very efficient in destroying parasites. A completely airtight building is required.

Pesticides can often be sprayed or dusted directly on to infested stock, providing directions are strictly adhered to, and convenient 6-oz aerosol sprays are available for domestic poultry-keepers wanting to treat individual birds.

Care needs to be taken to avoid chilling birds when they are sprayed, and the work should not be undertaken in cold or draughty conditions. It is unlikely that poultry would be treated just before slaughter, but if this is done, check carefully that no withdrawal period is needed for the pesticide in use. Sevin compounds, for example, are not to be used within a week of slaughter according to American recommendations, and the spray should not be in contact with nests or eggs.

Flying insects

Flies, certain moths and beetles are naturally attracted to any area where manure offers them a breeding ground and the right temperature and conditions, so the first rule in curtailing flying pests is to keep droppings as dry as possible and dispose of them some distance from both the poultry and domestic premises.

The ubiquitous house fly and its relations are a threat mainly through their disease-spreading habits of settling on faeces and feed containers in quick succession. Eggs are laid on manure or any putrefying material for the good reason that the larvae or maggots, which hatch in one to two days, eat the teeming bacteria around them. This is their contribution to the balance of nature. After a week to ten days they pupate, hatching into adult flies in a few more days.

Moths are less of a menace to poultry health, but in heavy concentrations can be a nuisance at house clean-out times.

Beetles are a mixed blessing. The small, black dung beetle *Carcinops*, for example, has a beneficial effect on droppings, aerating them and breaking them down into a dry and manageable material, whereas the meal-worm beetle *Alphitobius*, a wood-borer, can do considerable damage to the supporting structure and insulation of a building.

Many poultry keepers will opt for total destruction of all insects as the simplest solution of flying pests, but it is possible to be selective. Alphitobius beetles, for instance, are climbers, migrating into the upper levels of a house. This can be prevented by painting the walls of manure pits with diesel oil, or painting them and all supporting posts with creosote and an insecticide. Carcinops can be left to its own devices.

Dichlorvos (DDVP) is used in a number of proprietory insecticides, being available in strips of plastic which release an insect-killing 'vapour' into the air over a period of months, as well as in liquid solution for sprays and aerosol fogging, and in cartridges. It is active against all flying insects, including beetles. Diazinon is favoured by some makers and can be obtained in a 20 oz aerosol or as an insecticidal lacquer which can be painted on surfaces and remains lethal to crawling insects even after washing.

Pyrethrum, a widely used insecticide, is odourless and tasteless and can safely be used in the presence of livestock. It is frequently incorporated in oil-based products suitable for aerosol generators. And the list goes on, giving the poultry-keeper a large range of product costs and effectiveness from which to choose.

For general control, a residual insecticide can be coarse sprayed on to the outside of poultry houses, where flying insects may settle. Similar treatment can be applied to areas inside the house after first ensuring that nothing harmful comes within reach of the birds.

A clean, continuous kill of all flying insects is provided by electrically operated ultra-violet exterminators, which have the asset that no chemicals are involved. They lure the pests with ultra-violet light, electrocute them against a grid carrying a current that is safe to the human operator, then drop them into a tray below. The disadvantages are that a small running cost is incurred for electricity, and the range of the device is limited so that several may be required in a large house. Designs vary also, and I have come across some models in which the accumulation of dead insects was bridging the grid and causing short-circuits.

Animal pests

Among the traditional enemies of poultry, the fox must be the king. It is up to the poultry-keeper to see that all birds are secure at night, and if they get up before you do, that they can gain access to a safe run in the early morning, which often seems to be the time for foxes to strike.

Recently another marauder has appeared on the scene in England, however, the mink. Useful as disposer of dead chicks—hatcheries frequently have a contract to supply local mink farmers—these animals are a menace if running free. Under the Destructive Imported Animals Act 1932, the Mink (Keeping) Order 1972 requires occupiers to notify the Ministry of Agriculture of mink at large on their land.

Rats and mice have always been with us, and although rats will occasionally attack poultry, their main threat is as consumers and spoilers of feed and transmitters of disease. Any unguarded feed store is an open invitation to rodents, so feed should not be left out at night and all hoppers and containers should be as inaccessible as possible to raiders and resistant to gnawing.

Concrete floors will not stop rats and mice. They are excellent climbers. Downward facing metal collars can be fitted to the support piers of raised houses however. Anticoagulants are still the most widespread method of extermination, based on Warfarin, but unfortunately, in a poultry context, they tend to be limited in effect, not only because some animals have developed resistance to them, but also because Vitamin K, readily available in poultry rations, acts as an antidote to anticoagulants. Difenacoum is an alternative, as are the acute toxicants such as strychnine, but it goes without saying that poisons must be handled with special care—a good reason for calling in expert advice from the nearest Divisional Office of the Ministry of Agriculture.

Disorders

Apart from the major diseases of the viral and bacterial type dealt with in the next chapter, other factors such as stress, nutritional deficiency and accidents all play a part in the wellbeing of stock, and sometimes open the way for infections.

Stress

Practically anything out of the ordinary for the birds can lead to stress, travelling and arrival in strange surroundings being a classic example. Take no chances. Introduce new stock to clean, well ordered quarters at the optimum temperature, correctly lit, with adequate feed and water and they will at least start off on the right foot.

Stay with young birds for a time and see that they find their way around. Do not guess at the height of nipple drinker lines in rearing cages, for instance, but check that they really are easy for the birds to use.

Where there is any choice, always err on the side of generosity in terms of living space, trough allowances and water points. Some countries have tried to set statutory limits to cage sizes and stocking densities, but often these do little more than underline common-sense rules that the good stockman would apply anyway. In the UK, Codes of Animal Welfare were drawn up during the 1960s, based on the work of a government committee under the chairmanship of Professor Rogers Brambell. These quite wide-ranging codes lay down a number of recommendations, among them stocking densities for fowls and turkeys and the suggestion that they should have room to move, stand normally, turn round and stretch their wings.

The codes are not enforceable by law, although if a cruelty charge is brought they may be used in evidence, but research amply shows that the poultryman who crams in more birds to obtain more output is defeating his own object. Recent investigation with layers in America, using cages of 30×40 cm (12×16 in) and 30×46 cm (12×18 in) showed consistently better results, with lower mortality, lower feed consumption per dozen eggs and higher egg income over feed cost when birds were stocked three to a cage than when four were crowded in.

One stress problem which rapidly increases if birds are over-stocked is the concentration of ammonia. Birds will tolerate higher levels of ammonia than humans. Remember, however, that the gas is heavier than air, so that its density is greater near the floor, and therefore around the birds in deep litter houses, than at the head-height of the stockman.

While sensitivity varies between individuals, the average poultryman will not sufffer undue discomfort from ammonia at 15–20 parts per million for short periods, although the pungent odour and irritant effect of this gas on the eyes and nose is never pleasant.

Research with broilers has shown that appetite is affected at 40–50 ppm and above 50 ppm respiration is affected. Birds exposed to high concentration for long periods may contract ammonia blindness. Birds under severe ammonia stress stand with eyes closed, looking dejected. Often their back feathers will be damp and matted from rubbing their eyes which may appear bloodshot and swollen lidded.

Gas measuring devices are used by environmentalists to get precise readings but observation and experience should be enough for the poultryman. If you want to be more exact, litmus paper, impregnated with indicators in the pH range 6–11 can be obtained. Moistened with distilled water and held in the atmosphere for fifteen seconds it changes colour and can be compared with a scale to give a reading within ± 5 ppm.

The answer to ammonia stress is to check the birds frequently when low ventilation rates are being used. Keep stocking rates in proportion and litter dry, renewing it with each fresh flock.

Nutrient deficiencies

Dramatic symptoms, difficult to distinguish at first sight from infectious diseases, can be caused by a lack of one or more essential ingredients in a poultry diet although, bearing in mind the strides made by nutritionists, you may rightly wonder how poultry could suffer deficiencies on modern rations. The trouble is that 'receiving' does not necessarily mean 'eating', and deficiency problems can still arise for a variety of reasons.

Under intensive conditions a change in the environment, say a rise in temperature caused by a partial ventilation failure, will quickly reduce feed intake. Disease may depress consumption or interfere with the intake of nutrients. Feed may even be left in the trough because it is unattractive or unpalatable. Suddenly the carefully-balanced formula, designed for birds with normal appetites, becomes less than adequate.

Before this stage, careless handling or storage of finished feeds, or pressure and heat processes in compounding may have destroyed the vitamins. And sometimes the raw materials themselves are deficient. When the biotin level in many batches of wheat and maize was reduced in 1974, for instance, a disease known as Fatty Liver and Kidney Syndrome (FLKS) became prevalent among young birds. Addition of biotin to the finished compounds alleviated or prevented this.

Where imbalance creeps into a diet the first warning the poultryman is likely to have is an overall drop in performance. Here and there a bird may show an acute reaction. Such signs could point to innumerable diseases and disorders, of course, but an initial check through the symptoms which can arise from dietary deficiency could save a lot of unnecessary medication. The effects range a long way beyond the poor growth rate or actual weight loss normally associated with lack of food.

Poor hatchability

If eggs that fail to hatch are broken out and deformed embryos or dead-in-shell are found, a deficiency of *Vitamin A* among the parent birds may be the cause. This vitamin, or its precursor, carotene, comes from green foods, maize (corn) and carrots. Since these ingredients are not always included in compound rations, Vitamin A is frequently included as an additive in commercial foods. In domestic feeding, cod liver oil at 1%—less for very young birds—is a remedy.

If embryos have clubbed down, tiny nodules appearing in the forming plumage, a lack of *Vitamin B$_2$* (*Riboflavin*) is indicated. Again it is a vitamin which is generally added as a supplement, because it does not occur in sufficient quantities in regularly-used feed ingredients. In fact it is synthesized by bacteria in the gut of mature birds, and can be picked up by other birds where they have access to droppings, but this natural source is denied to caged birds.

Commercial compounds can be changed if a deficiency is found. Where feeds are home prepared, dried yeast and greenstuffs will counteract any shortage.

Among other hatchability problems the form of the embryo will often give a clue; parrot beaks, swollen heads and short down implying that breeding hens are receiving insufficient manganese.

Mucus discharges and skin eruptions

These symptoms can be the result of infective diseases, but *Vitamin A* deficiency is one avenue to check, because this vitamin helps to maintain the epithelial tissues and its absence leads to the nasal discharge and watering eyes known as roup.

Caseous (cheesy) material appears around the eyes and mouth and may block the bird's breathing. In the throat white pustules erupt, which are found at post mortem. Eye abnormalities and incoordination may occur. Turkey poults and ducklings exhibit similar symptoms to chicks. Geese on good grazing do not normally suffer the deficiency.

Scabs and crusting at the corners of the mouth, inflammation of the eyelids and dermatitis of other exposed skin areas will affect birds deprived of *pantothenic acid*, part of the Vitamin B complex. Wart-like swellings appearing under the toes when birds are about a fortnight old are another symptom. The vitamin is vital for proper growth and feather development. Again, dried yeast may be sprinkled on to home-prepared foods, fresh and dried grass also being good, but professional compounders add synthetic pantothenic acid to their products.

Severe inflammation of the mouth and tongue ('black tongue') seen in chicks after about fourteen days, suggests lack of another member of the Vitamin B complex, *niacin* (*nicotinic acid*).

Symptoms similar to those caused by a shortage of pantothenic acid, with roughening and cracking of the skin under the feet and scabby outbreaks around the beak and eyes can also result from a shortage of *biotin*.

Deformity

A common result of nutrient imbalance is deformity, either through softening of the bones, as in rickets, or by malfunction of the muscles and tendons, possibly accompanied by paralysis.

Vitamin D_3 is the important vitamin which ensures the correct assimilation of calcium and phosphorus by the body. Ultra-violet light acting on the tissues generates Vitamin D. Some is inherited by chicks from the breeding hen and some is to be found in properly adjusted diets, but its value is quickly lost if compounds are badly stored and allowed to become rancid. Without sunlight or Vitamin D in their diet, young birds suffer bone softening and swelling of the joints leading to the characteristic bowed legs of rickets. Those affected in maturity may show temporary paralysis and 'layers' cramp'. Cod liver oil included at up to 2% in chick rations is a recommended treatment for domestic flocks, but a dry supplement is normally added to commercial feeds.

A condition in which the leg-bones are bowed is occasionally found in ducks affected by lack of *niacin*. The bones however are rigid rather than soft as in rickets.

Another crippling condition which can arise is perosis or slipped tendon. Where there is too little *manganese* growing birds, particularly chicks and poults, suffer this disability. The dietary imbalance can be aggravated by an excess of bone meal. This contains calcium and phosphorus and tends to absorb manganese. The mineral is available in powdered form for compounders, but the outer husk of wheat germ (millers' offals) is also rich in manganese.

With perosis the hock joints appear swollen and the tendons slip out of position so that the lower legs splay outwards, making it difficult or impossible for the bird to stand or walk. Depressed levels of *choline*, *biotin*, *niacin* and *folic acid* also contribute to perosis. All these vitamins are found in normal poultry feed ingredients: choline in cereals, protein meals and milk products; biotin notably in brewers' yeast; niacin also in yeast, milk products and greenstuffs and folic acid from green leaves, hence the name, as well as yeast and liver meal. But they can easily be 'lost' by bad handling.

Some deformities are caused mechanically rather than by actual dietary shortage. Shovel beak or mandibular disease comes into this category. Feed texture is the culprit, fine dusty rations clogging the beak and mucous glands and causing degeneration of the mandible. Wet mash, then the use of a coarse ground feed, may solve this problem.

Nervous abnormalities

Paralysis and nervous spasms are often connected with severe nutritional deficiencies, the classic being 'crazy chick disease', resulting from a lack of *Vitamin E*. Brain damage or encephalomalacia causes the young birds to weave about and fall over, paddling their legs in the air. Once again feed handling and storage is the key. The vitamin is present in wheat or maize germ oil, green foods and fresh animal fat, but it is readily destroyed by rancidifying fat, so feeds must be used promptly and not stockpiled where damage can be done.

Curled toe paralysis in chicks and poults comes from a shortage of *Vitamin B$_2$*, mentioned earlier for its effect on hatchability. The toes curve round until the bird cannot stand, and instead squats on its hocks, shuffling reluctantly on the outer edges of its legs. It culminates in paralysis. As birds mature, however, their dependence on dietary Vitamin B diminishes and they may literally 'grow out' of these symptoms.

Fatty Liver and Kidney Syndrome has already been mentioned as the outcome of low *biotin* levels in a cereal harvest. It has caused serious economic losses among broilers particularly, the birds becoming paralysed and prostrate, with their necks outstretched. As much as a fifth of a flock may die. The susceptible age is eleven to thirty days of age, and although less obviously defined than some deficiencies, the biotin connection seems clear. Where a high proportion of wheat is involved in a diet, compounders now check whether biotin supplementation is necessary.

Lack of *manganese*, referred to previously as a main cause of perosis, can also lead to head retraction in newly hatched chicks from manganese-deficient parents. They frequently recover in a few hours however.

Internal haemorrhages and possibly bleeding to death from minor injuries among chicks may signal that their parents were short of Vitamin K. This is normally sufficient in fresh green foods for domestic flocks, and can be incorporated in synthetic form in commercial feeds, but its uptake by the body may be inhibited by sulphonamides and tetracycline used as anti-coccidial and antibiotic treatments respectively.

Poisoning

Occasionally trouble stems from the presence rather than the absence of a feed ingredient and the reader should be alert to cases of possible poisoning where birds have been exposed to cresol disinfectants, nicotine sulphate used as an insecticide or any other 'unusual' compounds shortly before showing symptoms. Zinc phosphide and arsenic, sometimes employed to kill vermin, can be fatal to birds.

An unpleasant, but fortunately rare, form of poisoning is caused by the organism *Clostridium botulinum*. This is botulism and it arises when birds peck at putrefying matter containing the organism, maggots or other victims of the toxin. Ducks in an open environment may be at greatest risk, but intensively housed stock, such as broilers, can also be poisoned if stockmanship is lax and dead birds are left on the floor for prolonged periods.

Botulism has the effect of paralysing the neck muscles, leading to a condition known as limber-neck in which the bird cannot hold up its head and eventually rests it on the ground. Little, short of culling, can be done for sufferers, but thorough disinfection of the site will be needed.

Although *aspergillosis* is not, in the conventional sense, a symptom of poisoning (in fact it shows up as a respiratory problem) it is sufficiently closely connected with feed to be mentioned here. It is caused by a fungus, *Aspergillus fumigatus*, which affects many species and can be dangerous to man.

Mouldy hay is the most generally recognized source of the fungus, but any environment where dust, stale feed and humidity abound can lead to outbreaks. A prime cause of moulds, for instance, is the addition of wet mash to feed troughs without cleaning out stale left-overs from a previous feed.

When the fungus affects young birds it is often called *brooder pneumonia*, being accompanied by respiratory distress and rapid breathing but none of the rattling or gurgling associated with other respiratory diseases. Diagnosis can frequently be blurred by stress from other conditions, however, such as pullorum, described in the next chapter. Sometimes it affects mature birds.

Post mortem investigation shows lesions in the lungs and air sacs, where the mould may be identified as a thick, furry growth. No practical treatment for victims is known.

The spores can also affect hatching eggs, penetrating the shell and colonizing the air-space with green mould. Dirty eggs should not be used for hatching, even if washed, because water helps to transmit the spores, but fumigation with formaldehyde is effective.

Aspergillosis is a good example of *mycosis*, which is the growth of a fungus directly in the body. Another condition caused in this way is *moniliasis*, which generally affects poults between two and six weeks of age. Here the causative agent is *Candida albicans*, which reduces growth and causes poor feathering and listlessness. On autopsy the crop lining is found to be ulcerated and raised in a rough 'pile' like towelling. The mouth can also be affected and the disorder is commonly known as thrush. The best remedy for this problem seems to be the maintenance of good management and hygiene, which should prevent it.

The alternative to mycosis is *mycotoxicosis*, which is a long way of saying that the fungi contaminate the food rather than making a direct attack on the body. Fungi in this category are collectively called myco-toxins.

Probably the best known example of poisoning by a mycotoxin is *afla-toxicosis*. Extensive detective work narrowed it down to ground-nut meal contaminated with a toxin caused by *Aspergillus flavus*. Birds poisoned with it become listless, lose appetite and die in about two weeks. Duck-lings are particularly susceptible.

Modern compounders, alert to the danger, have greatly reduced the threat by screening their feeds. Nevertheless there are at least 200,000 species of fungi in existence. Most are harmless and some, like yeast, of positive value, but the spores are ever present, and so is the risk of mycosis or mycotoxicosis. Conditions favouring mustiness, with feed moisture levels above about 13%, relative humidity around 80% and ambient temperature above 13°C (55°F) present the greatest danger.

Some control can be obtained by spraying finished feeds with pro-pionic acid which inhibits the spread of fungus, but the main factor is to keep storage bins clean and dry, troughs clean and drinkers properly adjusted to avoid unnecessary spillage.

First aid

Occasionally birds suffer ailments calling for first aid. In layers the prob-lems are often associated with the reproductive tract.

Egg binding

The inability of a hen to produce an egg when she is making obvious attempts to lay, with frequent visits to the nest and signs of distress, suggests an abnormally large egg or a contraction of the oviduct. Lift her carefully, since an egg broken internally can be disastrous for her. Warmth and humidity are best for relaxing the duct. Hold the vent of the bird over a jug of hot water to which iodine has been added at the rate of ten drops to 1 litre (2 pt). Alternatively, lubricate the vent with a little olive oil. Then replace the bird in her laying quarters. If these remedies fail, something more serious must be suspected. Normal-sized fowls have been known to lay eggs weighing as much as 285 g (10 oz), five times larger than average, without coming to harm.

Internal laying

Where a hen is totally unable to lay, the problem is likely to be internal laying. The maturing egg, failing to be collected by the infundibulum, passes instead into the peritoneal cavity where it may become infected, leading to egg peritonitis. Little can be done for a bird in this state, which will die if not put out of its misery.

Prolapse

Sometimes, after much straining, a laying bird will succeed in producing an egg, but at the expense of forcing out part of the oviduct, which protrudes through the cloaca. Older birds are more subject to this than pullets. It needs to be corrected fast, not only for the comfort of the bird but also because the red protrusion is a focal point for other birds, who will peck at it if they have the chance, causing real damage and creating the vice of cannibalism which can be very hard to eliminate in a flock.

Supporting the bird head down on the lap or on a table, wash the organ gently in warm water, apply some olive oil and work it back inside with the finger.

Prolapse may happen again, in which case the bird is probably best for the pot, but with luck the sufferer can stage a complete recovery. Isolate her for a few days from the rest of the flock.

Cloacitis

A discharge of milky-white, offensive matter from the vent can be caused by inflammation of the mucous membrane of the cloaca, and is known as cloacitis or 'vent gleet'. It is thought to be venereal and appears to be contagious. Affected birds should be removed from the flock; if they are forced out of production, and the vents painted with iodine, some of these birds will recover. Otherwise they should be culled.

Cannibalism

More of a habitual trait than a physical ailment, feather-pecking and cannibalism can cause many difficulties for the poultry-keeper. It has become common practice on many commercial units to have birds beak-trimmed or 'debeaked' at an early age to prevent the vice developing. A secondary reason put forward is that this cuts down food wastage by stopping the birds from flicking feed. Looked at from a strictly commercial angle, these arguments carry some weight, and it can be said that modern, lightweight, high performance birds, being highly strung, call for special treatment. Beak trimming is condoned by welfare authorities, but making it routine can easily be used as a cloak for bad management. Birds may have parasites which cause feather-pulling, for example. To inhibit proper preening by beak trimming in those circumstances may lead to more misery for the birds, simply because the stockman failed to take proper hygiene precautions.

Vice is not purely a 'big flock' problem. It can, and does, occur on range and is variously attributed to boredom, curiosity, excessive light and stress. The bird attacked often appears resigned, hunching up and accepting its fateful position in the peck order without resistance.

Prolapse has been mentioned as a starting point for some outbreaks, but normal birds immediately after lay also temporarily display a pink area which can induce others to peck at the vent, and this is a good reason for keeping nest boxes properly darkened. Shades and/or painted windows are used to subdue the light in intensive houses.

Ensuring that birds are not over-crowded, which might lead to accidental treading and thence to cannibalism, and that they have sufficient trough and drinker space are elementary ways to keep down stress and so, possibly, avoid worse problems.

At one time blinkers, fitted like spectacles over the beak to prevent the bird from seeing what it was pecking, were fairly widely employed, but the cost of purchasing and fitting were against them, and although they are permissible in the UK, legislation lays down that such devices must not penetrate or mutilate the nasal septum. They are rarely used now. Debeaking may be the only answer if trouble persists.

Where victims are spotted early enough, they can be isolated from the main flock and the wounds treated, an old remedy being one part creosote to thirty parts lard or petroleum jelly.

Egg-eating

This problem can be set off by an egg becoming cracked, leaking its contents and the birds acquiring a taste for it. If offenders can be traced, they may be deterred by putting in decoy eggs which have been hard-boiled, carefully opened and filled with mustard or, more elaborately, with a paste made of flour and a little ginger and asafoetida in equal proportions.

Bumble foot

A graphically named disability, bumble foot is most commonly a problem with heavier birds subjected to wire floors or narrow perches. Sometimes landing on a hard floor can cause it.

As the result of bacterial infection in a minor injury such as a cut or puncture, the foot suffers a ball-shaped swelling underneath, or between the toes, which is really an abscess. Treatment is to bathe the foot in warm water, cut a cross in the head of the swelling and squeeze out the pus. Then clean the wound with water disinfected with a few crystals of permanganate of potash and bind it with a sterile bandage. A veterinary specialist may use an antibiotic such as penicillin or tetracycline to aid recovery. Surfaces on which birds stand should be smooth and free from sharp edges.

Crop impaction

String, long grass and, in recent years, polythene bags, are among the many strange things which poultry are liable to pick up, particularly on range, but occasionally it brings them trouble in the form of crop impaction.

This causes a distended crop, even when the bird has not been fed for a day. It will feel hard to the touch. Use a syringe, if available, to administer about half a tumbler of warm water to a chicken in this state, then upend it and gently massage the crop to expel the contents. If this fails and the bird is valuable, an incision of about 6.3 cm ($1\frac{1}{2}$ in) can be made in the outer skin and a second of the same length through the crop wall at right angles to the first, to open the crop, but this is really a job for a vet. After suturing the cut, the bird must have only soft food in small quantities for a week. Otherwise it should be humanely destroyed.

A similar condition, *sour crop*, causes distension but the crop feels soft. Massaging can again be applied, frequently with success. Withholding feed and limiting water for forty-eight hours and administering small amounts of alcohol (brandy will do) can sometimes be used to advantage.

12 Major infective diseases

There are several good reasons for having at least a working knowledge of the major diseases. Firstly, if the worst happens and something nasty crops up, an accurate report to a veterinarian, based on a knowledge of what to look for, can save a lot of time in reaching a diagnosis. It can also help if you know what to do before medical assistance arrives, and, certainly with increasing experience, vaccines and treatments can be given by the poultryman unaided. There is another point. Understanding everyday terms used in the industry can sometimes save a red face.

Marek's disease

Although Professor Marek, a Hungarian veterinary pathologist, described the disease named after him in the early 1900s, it was not until the mid 1960s that a vaccine was developed to prevent it. A prime reason for this was that from an early stage Marek's disease was regarded as part of the avian leukosis complex. In a way this is understandable, because Marek's and the leukosis diseases cause tumours which can be hard to differentiate, but they are, in fact, separate conditions.

Leukosis is found in different forms, lymphoid leukosis being the most common, followed by myeloid leukosis and erythroid leukosis. It is caused by a group of RNA tumour viruses called the avian leukosis/sarcome group. Erythroid leukosis has a leukaemic effect on the blood, the other forms leading to tumours in the major internal organs.

Marek's disease, officially named neuro-lymphomatosis gallinarum, arises from a cell-associated DNA virus of the herpes virus group.

Because of the long-standing relationship of Marek's with leukosis there are no early statistics to say what damage the individual diseases inflict, but it is now recognized that Marek's alone, prior to the development of a vaccine, was the single most potent cause of death among fowls. The breakthrough to its control came when it was separated from the pack by British scientists and the herpes virus identified.

Outwardly, Marek's disease appears in chickens from six weeks of age onwards. Frequently birds can be found in a characteristic 'straddled' position, one leg forward, the other back, gripped by progressive paralysis of the legs and wings. In an unvaccinated flock the spread can be rapid and the incidence of death easily 25–30% or more from an acute attack. Lymphoid leukosis, on the other hand, is not seen in birds under sixteen weeks of age, is much less likely to be accompanied by paralysis, is not particularly infectious and mortality is low.

Classical, as opposed to acute, Marek's may kill 10–15% of a flock, symptoms showing that the peripheral nerves serving the limbs, the respiratory system and the digestive tract, for example, have been affected. Internally, diseased nerves are often greatly enlarged and sometimes small tumours are detected in the main organs.

Birds may die suddenly with no symptoms from acute Marek's. Tumours in, and gross enlargements of, organs such as the liver, kidneys, gonads, lungs and heart are commonly found on autopsy.

Dust is the great enemy where Marek's is concerned. The highly infectious virus particles are present in feather follicles as well as at the mouth and cloaca of infected birds, and travel easily on dust arising naturally. Once established in a house, dust-borne virus can remain infective for a year.

No treatment is known for diseased birds, which emphasizes the importance of immunization. Before this became possible breeders concentrated on producing resistant strains of layers, and this work was not wasted. Even with vaccination, it still pays to have stock that stand up well to the disease.

In the early stages, field strains of Marek's virus affecting chickens were modified to provide suitable vaccines, but at much the same time a herpes virus in turkeys which produced a mild reaction in both turkeys and chickens was studied, and vaccines developed from it were found to give chickens good protection against the disease. Most vaccines are now based on this turkey herpes virus.

For convenience, the vaccines are widely prepared in 'lyophilized' (freeze-dried) form, although 'wet' varieties are also available. They are administered by a single intra-muscular injection, usually into the thigh of the newly hatched chick, so this is a hatchery routine.

The strength of a Marek's vaccine is measured in plaque forming units or pfu's and the minimum recommended dose is 1,000 pfu's.

Generally, vaccination is reserved for pullets, although some producers apply it to broilers and claim to have obtained better results from their stock.

Tumorous conditions recorded in turkeys and other species are occasionally associated with Marek's and more frequently with leukosis, but the economic importance of these outbreaks is infinitesimal compared with the losses once common among commercial layers. Leukosis diseases are egg transmissible, progeny sometimes inheriting infection from parents, but being relatively non-contagious they can be controlled by eliminating known carriers, keeping young birds carefully segregated from older stock and putting them in thoroughly disinfected premises. Bird strains vary genetically in their susceptibility to the various forms of leukosis.

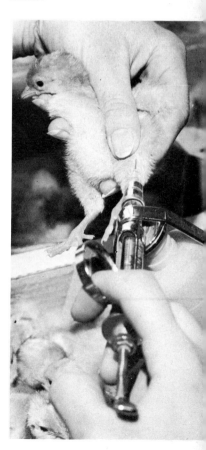

Marek's vaccination at day-old has dramatically reduced losses from this disease

Newcastle disease

Everybody, even those not connected with poultry, has heard of fowl pest. More correctly known as Newcastle disease, because 'fowl pest' also includes fowl plague, a different infection, it is a 'notifiable' disease and for a number of years in the UK notification to the authorities meant compulsory slaughter of all birds and the payment of compensation by the Government. The advent of satisfactory vaccines in the early 1960s changed that and today slaughter with compensation is discretionary, only likely to be applied in the case of a severe epidemic.

The disease gained its internationally recognized name in the 1920s, after outbreaks had been identified in the area of Newcastle upon Tyne. Caused by a myxovirus belonging to the para-influenza group, it affects a wide range of birds, including turkeys, and can be carried by geese, ducks and wild birds which may themselves show no symptoms. In its most virulent form it can wipe out an entire flock.

Birds can pick up the virus through the eye, as well as via the respiratory tract, and the early signs of trouble are an increase in water consumption, loss of appetite and respiratory difficulties marked by coughs, sneezes and a high-pitched rattling sound. The neck may be extended and the beak opened wide. In laying flocks egg production drops dramatically. Depression sets in and the faeces often appear loose and green. Later, nervous symptoms occur: sudden involuntary movements, a stiffening of the gait, followed by paralysis of one or both legs and possibly of the wings and neck. Eggs, if laid, are often thin-shelled, misshapen and lacking in pigment.

Strains of Newcastle disease are grouped in three classes according to their virulence for fowls: velogenic, mesogenic and lentogenic strains. Velogenic viscerotropic Newcastle disease (VVND) is the cause of many of the more severe outbreaks recorded.

Successful 'inactivated' vaccines were developed first, which had to be injected into each individual bird, but a second form of vaccine, based on attenuated live strains of the virus, which could simply be added to the drinking water, was quickly adopted by poultry-keepers.

Live ND vaccines are generally prepared with lentogenic strains such as Hitchner B_I and LaSota (named after the people who first isolated them). Mesogenic strains of virus such as Roakin and Komarov, although used occasionally for revaccination and at sites where no ND susceptible birds are kept, are too pathogenic for primary vaccination.

LaSota has proved particularly useful for future layers and breeding birds because, having a higher potency than Hitchner B_I, it gives protection over a longer period and cuts out the need for frequent revaccination during the laying period. Its disadvantage is that it is apt to depress growth and food conversion if used on very young birds and therefore is unsuitable for broilers.

Vaccination schemes are frequently based on a combination of Hitchner B_I and LaSota, the milder Hitchner strain being used first, followed by a sequence of LaSota doses to consolidate immunity.

These vaccines may be applied in several ways, depending on circumstances. Where areas of high ND infection are known to exist, chicks may be sprayed with Hitchner B_1 on arrival at the brooder house, before they are removed from their boxes. Sometimes the vaccine is combined with an Infectious Bronchitis vaccine. Subsequent LaSota doses may be by nasal or eye drops, but the most common method is by simple addition to the drinking water. In the USA it is sometimes applied as a dust in totally enclosed houses or fogged in as a mist; water application, while still used by fanciers, is loosing favour with commercial growers. A golden rule, of course, is to read the vaccine manufacturer's instructions with care and to follow veterinary advice.

Infectious bronchitis

Symptoms of infectious bronchitis can, at first, be easily confused with ND. The birds develop a high-pitched cough and make a rattling sound in the throat. Weakness and depression set in and nasal discharge and wet eyes are common. Droppings are loose and greenish in colour. In contrast to ND however, egg production shows no immediate pronounced drop.

Among poultry this disease occurs only in fowls and is extremely contagious. The virus can be transmitted directly from bird to bird or via the air, drinking water, clothes of the stockman, and so on.

Effects of IB may appear within eighteen to thirty-six hours of the birds becoming infected. The initial symptoms pass off in about a week if they are not aggravated by complications such as *Mycoplasma gallisepticum* or *E. coli*, and except where chicks are involved, mortality is usually quite low. But this is not the end of the poultryman's problems.

Within a few days of disease onset, egg production declines quickly, often staying down for weeks or even months. Eggs which are produced have rough, wrinkled or oddly shaped shells. The albumen becomes thin and watery. Some eggs may be soft or laid without shells, and internal laying can affect a proportion of the hens. Young pullets do not escape these effects by contracting the disease before coming into lay. In fact the younger the birds, the higher the percentage likely to be non-starters in production. Development of the oviduct is inhibited or stopped, while general body growth is liable to be stunted and feed conversion totally uneconomic. Frequently the way is opened for attacks of chronic respiratory disease (CRD) and other infections.

Several serological types of IB virus have been isolated, but the majority of outbreaks of the disease are caused by the Massachusetts type. Live vaccines prepared from this virus induce good protection against field infections of both Massachusetts and a less pathogenic form of the disease, the Connecticut type.

The vaccine is cultured in eggs and has to be attenuated to prevent unfavourable reaction in the birds. This is achieved by 'passaging' it a number of times through eggs, and the strength of the inoculum is often expressed by the number of passages. An H-120 vaccine strain, in other words, is attenuated by 120 egg passages to provide a mild initial vaccination. This can then be followed by a stronger H-52 strain when the birds have developed some resistance.

Several pharmaceutical companies manufacture IB vaccines and a variety of routines are put forward. Where infection is already present on a site, spray vaccination of chicks within their first five days is generally recommended, and early vaccination is also widely practised with broilers.

If the number of birds is not too great, vaccination may be by instillation. Using a dropper, a single drop of vaccine is introduced to the nostril or eye. The method results in a good, even reaction through the flock, but clearly takes longer than mass vaccination. It is best suited to chicks, but can be applied to mature birds. Beak dipping is an alternative.

The combination of IB and ND vaccines has proved quite successful and has the advantage of reducing time, labour and the number of stress reactions for birds. If the vaccines are used separately, however, they must be applied with at least two weeks between them.

Layers and breeders not under serious challenge from IB will normally receive their initial mild vaccine in the drinking water at three to five weeks of age. Slight respiratory symptoms may follow after four or five days and persist for up to a fortnight. At fourteen to sixteen weeks a second water treatment using the stronger vaccine is given. Usually this causes no reaction.

As an additional precaution with layers, a further vaccination is sometimes given at moulting time.

Infectious laryngotracheitis

Occasionally it becomes necessary to vaccinate for another disease which causes respiratory symptoms, gaping and rasping in the throat. This is infectious laryngotracheitis (ILT). The disease leads to quite severe damage of the lining of the larynx and trachea and birds frequently cough up blood, but it is a more sporadic infection, with less spread, than the other major respiratory diseases. For this reason vaccination is only employed when there is definite evidence of a challenge in a group of birds, or other outbreaks have occurred in the area. In fact it is one of the few virus diseases where vaccination can still be effective in a flock even after some birds have shown symptoms.

Caused by a Herpes group virus, ILT only seriously affects chickens. Turkeys, ducks and geese appear resistant. It may result in high mortality where the peracute form is involved, birds having obvious difficulty in breathing and dying of asphyxiation within two or three days. In the subacute and chronic variations of the disease mortality is lower, symptoms extend over a longer period and only a few birds may be affected.

If ILT is suspected, veterinary advice should be sought quickly. The only successful method of vaccination is by eye drops. Spraying or drinking water techniques do not work and simply represent a waste of vaccine.

Since the disease is easily transmitted on clothes, equipment, vehicles or contaminated feed, strict hygiene is needed at all levels. Footbaths containing a 3% Cresol or 1% lye solution should be properly used. Transmission by air is possible, particularly on multi-age sites.

Providing there are no complicating factors like coryza or mycoplasma infection, birds can recover from ILT in ten to fourteen days, but they can become carriers and a long-term threat to other susceptible stock. Sometimes the only answer is to dispose of an infected flock.

Fowl pox is easily recognized by the lesions it causes

Fowl pox and diphtheria

Although fowl pox and avian diphtheria in chickens show quite different clinical effects, they are actually symptoms of the same virus. The disease is widespread and attacks a variety of wild and domesticated birds, including the pigeon, fowls and turkeys. It is now well controlled by vaccination.

Pox is not difficult to identify. After a day or so when the birds become depressed, lose appetite and may start to pass a greenish diarrhoea, small grey eruptions appear on any unfeathered parts, particularly the comb and wattles, eyelids, legs, between the toes, and around the beak and cloaca.

In a short time the eruptions turn yellow and then encrust into wart-like spots. Often the eyes water, become inflamed and may be permanently damaged.

The diphtheritic form results in cheesy yellow membranes in the mouth, throat and trachea. Food swallowing becomes difficult and badly affected victims die of suffocation. Mortality is not such a problem as the economic effect on meat or egg production however, since the spread of the disease is usually slow. It enters the bird either through tiny wounds or scratches on the skin, or via biting insects which act as carriers. The virus itself can survive for several years in scab tissue from the pox lesions.

Once an outbreak is diagnosed, all healthy birds should be vaccinated without delay using the follicular method described on page 179.

Avian encephalomyelitis

Epidemic tremor is the common name for AE, a highly contagious disease mainly affecting young birds and reported in chickens, turkeys, quail and pheasants. Ducklings and guinea-fowl are also susceptible. Due to its economic importance many poultry-breeders now include AE vaccination in their routine immunization programme.

The picornavirus responsible for the disease does much of its damage in the hatching egg. If a laying bird having no immunity is exposed to infection, the virus can reach the eggs via the ovary and may paralyse the developing chick sufficiently to prevent it breaking from the shell, with the result that many dead-in-shell are found.

Infected chicks that succeed in hatching generally start to show difficulty in walking and may develop a tremor of the head, the neck and occasionally the whole body within about a week. Progress of the disease ends in total paralysis.

Mortality can be very high among susceptible birds, but sometimes only a few are affected. Apart from embryonic infection, AE can also be spread by entering the mouth and alimentary canal, so cross-infection between chicks is quite possible. Survivors may suffer opacity of the lens of the eye and sometimes total blindness.

Although the disease is most dramatic in young birds, it can affect all ages. Mature birds not in lay show no recognizable symptoms. Those in lay which are susceptible may demonstrate a sudden drop in egg production, and if they are breeders hatchability will be impaired. Because of this, and also because of the implications for future stock, vaccination of breeding stock is clearly important. Layers less frequently receive this vaccination, which must be slotted into a programme at least a month before the onset of lay and two weeks away from other vaccinations.

Live and inactivated AE vaccines are available, the live version being commonly applied in routine immunization. It can be placed in the drinking water or introduced by oral instillation. By this method a limited number of birds, at least twenty in a flock of 500, is given the vaccine solution by mouth from a dropper bottle. The virus then spreads through the flock, conferring immunity for a laying season. In circumstances where birds in lay must be treated, the killed vaccine may be used.

Infectious bursitis

The disease of infectious bursitis is popularly known as Gumboro disease because the first outbreaks were observed on farms around Gumboro, in Delaware, USA.

Since it was reported, in 1962, veterinary scientists have been intrigued about its cause. Several virus types have been suspected, the latest being a diplorna virus, but there is still some speculation about the precise nature of the 'infectious bursal agent', although effective vaccines have now been developed against it.

To the poultry-keeper it is seen as a short-term illness, coming on rapidly. At first birds may develop a slight tremor of the head and body. Within a day or so appetite has dropped, depression has set in and vent feathers show staining with white, watery diarrhoea. Vent picking may be observed, and the worst cases become prostrate and die.

Gumboro usually occurs between eleven and thirty-five days of age, so growing pullets and broilers are commonly the victims, although it has attacked birds up to fifteen weeks of age. Only fowls appear to suffer from it. Mortality can be 20% or more although it is often less than 10%. Its effects may be seen in the entire flock. Recovery is rapid, usually within a week.

While the direct effects of Gumboro are bad enough, its real economic importance is the influence it has on resistance to other diseases. This is because it concentrates principally on the Bursa of Fabricius, the small organ described in Chapter 2 which plays such an important part in antibody production. The Bursa swells and becomes oedematous and yellowish. In severe cases haemorrhage and necrosis occur. It may then regress to less than normal size.

Bursal damage leads to immunosuppression, inhibits the benefit a bird can obtain from vaccines, notably those for infectious bronchitis and Newcastle disease, and could account for some vaccine failures. At the same time, reaction to live respiratory vaccines is likely to be more severe in immunosuppressed birds.

With the vaccine procedure now worked out, using freeze-dried vaccines, the immunity status of the flock is important. If the birds are susceptible and have no maternal antibodies to the disease, they must be vaccinated as soon as possible, and the vaccine can be combined with Marek's vaccine, injected intramuscularly at one day old. The vaccine will not 'take' in the presence of maternal antibodies, however, and where there is any doubt, birds must be re-vaccinated at thirty days of age, when their natural immunity has waned. At this stage the vaccine can be administered by spray. It combines effectively with Hitchner B_1 Newcastle disease or H.120 infectious bronchitis vaccines. Alternatively, parent flocks may be vaccinated so that parental immunity is passed on to the next generation, which then only needs the vaccination at thirty days.

Adenoviruses

In the mid-1970s the adenoviruses started to be implicated in several conditions appearing in poultry, notably sudden drops in egg production.

First isolated in human adenoids—hence the name—adenoviruses form a large group of viruses found in animals and birds. There are at least eleven serotypes or species of fowl adenovirus (FAV), and the most widely recognized is FAV-1 CELO (which stands for chick embryo lethal orphan). Another is classified as the Tipton strain. In the autumn of 1976, however, a new syndrome was identified in Northern Ireland by Dr J. B. McFerran and classified as adenovirus '127', while Dr W. Baxendale of Intervet Laboratories Ltd, in England, identified strain 'BC 14'. The work of both researchers was an important step in tracking down unexplained falls in egg output, now known as the egg drop syndrome or EDS 76.

Egg drop syndrome

This can easily be confused with IB in its effect on layers. In fact early reports were put down to failures in IB vaccination. Birds show a rapid fall of 10–40% in egg production and eggs with rough, thin and soft shells, often lacking pigment are laid. Failure to reach peak performance is a common characteristic. Sometimes severe diarrhoea may be seen, but this is not consistent, and apart from a loss of appetite, birds appear healthy.

Broiler parent flocks were the first to show signs of the disease but it has become widespread among commercial layers in Holland, where bad internal egg quality has also been reported.

Early in 1978 a vaccine against EDS 76 from Intervet Laboratories Ltd, was licensed in the UK by the Ministry of Agriculture. It is given once only as a subcutaneous injection at fourteen to twenty weeks of age.

Inclusion body hepatitis

This is caused by the Tipton strain adenovirus referred to above and affects broilers and layer replacements. It is marked by a sudden increase of deaths at around five to seven weeks of age. Few outward signs of illness may appear, but up to 10% may die, while the rest of the flock appear healthy. Liveweight in broilers may be depressed. Internally, the liver and kidneys are enlarged and haemorrhaged, and other internal organs may be involved.

Duck diseases

A number of virus diseases, such as Newcastle and ornithosis, are common to most poultry species including ducks, but two important infections are particularly associated with ducks.

Duck virus hepatitis

This causes high mortality among birds up to three weeks old. Death is rapid. Victims fall over, paddle aimlessly with their legs and frequently die within an hour. The head is drawn up and back in spasm. In susceptible flocks as many as 90% may be lost. Internally, the liver is found to be severely haemorrhaged.

A vaccine has been developed which is either given at day old by stabbing the foot-web or, where birds are known to have inherited maternal antibodies, by injection at ten days of age.

The virus survives well in dirty conditions and probably becomes active when ingested, so control by strict hygiene can help.

Duck virus enteritis

Otherwise known as duck plague, this was not seen in the UK until 1972, when it occurred among ornamental ducks although it was first diagnosed in Holland in 1949. It is also found in the USA. This enteric disease may be noticed first when birds become mopey and reluctant to move. Mortality may be very high, but a live vaccine is available which is effective in the face of a diagnosed outbreak.

Use of vaccines

Obviously, whenever serious illness strikes a flock, professional advice should be sought. But it should not be necessary for anybody rearing birds to need expert help with routine vaccination once the initial programme has been laid down and the type of vaccines have been established. On the other hand, certain basic principles underlying vaccination need to be clearly grasped if time and money are not to be wasted.

It should not be hard to trace vaccine suppliers. Companies making vaccines frequently advertise in poultry magazines, and the *Disease Directory* published each April and distributed internationally with the journals *Poultry World* and *Poultry Industry* contains a full list of manufacturers and their health products. Government officials (local ADAS offices in England and Wales) are always ready to give direct assistance or will recommend suitable diagnostic services and veterinary surgeons.

Having decided on a certain programme, however, stick to the instructions.

Timing is of great importance with vaccination, because the level of antibodies, the body's defence mechanism, varies with age and the type of disease-challenge experienced. Because resistance takes time to build up, immunity against certain diseases is inherited by the young as antibodies from the mother. This maternal or 'passive' immunity is gradually replaced by the antibodies of the individual, but if a vaccine is used before the parental antibodies have declined, they may fight off the new vaccine and neutralize its value. In some diseases, such as avian encephalomyelitis, maternal immunity can last for many weeks, while for others, such as ND and IB, it is not so prolonged.

Fit birds, well fed, hygienically housed and stress-free, generate their own antibodies and more effective 'active' immunity after vaccination than sickly ones. As far as possible, therefore, birds should be built up for vaccinations, and this is an occasion when extra vitamins and minerals can be useful.

Storage of vaccines according to company recommendations has a direct bearing on their efficiency. Many vaccines are lyophilized and sealed into small glass vials. These should be stored in darkness at a temperature of 4° C (39° F), as in a domestic refrigerator, and the expiry date on the label usually assumes this has been done.

Vaccination methods

The drinking water system

The easiest method of vaccination, particularly when large numbers of birds are involved, is to introduce vaccine to the drinking water. This

is suitable for birds of two weeks old or more, and can be applied with
ND, IB, ILT and AE vaccines. Here are the rules:

Use buckets or containers which you are sure are free of all contaminants, such as disinfectants or detergents. If the vaccine is
to go directly into header tanks, the system must be clean.

Make sure the birds will drink readily, if necessary by depriving
them of water for two hours beforehand.

Water must be cool, non-chlorinated tap or rain-water with no iron
in it.

Having checked dilution instructions, open the vials under the
water and mix thoroughly, using a clean stirrer.

Do not expose vaccine-medicated water to direct sunlight.

Regard the dosages quoted on packs as a minimum and if the
number of birds falls between recommended levels, choose the next
higher dosage. Try to ensure that all the vaccine is taken up within
two hours. The addition of milk in equal parts with the water increases the active life of the vaccine.

Spray systems

Another method of mass vaccination with live virus vaccines is by spray.
Infectious bronchitis and Newcastle disease vaccines are occasionally
applied in this way to day old chicks in areas of high-infection risk.
Normal procedure is to spray young layer or broiler stock while they
are still in their boxes on arrival at the rearing quarters.

Pump-up pressure sprays of the type used by gardeners are often used
but because the size of spray particles is critical, an adjustable nozzle
is important. Specially made sprayers are obtainable from pharmaceutical suppliers. Sufficient vaccine for 1,000 doses can be mixed in
1.5 litres ($2\frac{1}{2}$ pt) of water, which must be pure or distilled and at a
temperature of 25°C (77°F).

It is possible for humans to suffer conjunctivitis from these sprayed
vaccines, so an eye mask should be worn when testing and using the
apparatus. Adjust the nozzle to obtain a medium density spray of even-
sized droplets. Wet the birds lightly by spraying from a height of about
60 cm (2 ft) and allow them about fifteen minutes in the boxes to dry
in the natural warmth of the room before putting them under brooders.
Never dry them under the brooders.

This system protects chicks against IB and ND independently of
maternal antibodies for up to four weeks. At three weeks they need to
be vaccinated by the drinking water method to build up lasting immunity. For the spray technique to be successsful it is important that
the birds should be healthy and treated within less than thirty-six hours
of hatching.

Instillation

More laborious than the drinking water method, but ensuring every bird
is dosed, the intra-ocular or intranasal routine can also be used in an

emergency on chicks less than two weeks old. The equivalent of 500 doses of vaccine can be dissolved in 15 ml ($\frac{1}{2}$ fl oz) of saline solution. From a dropper bottle, at a height of a few centimetres (up to 3 in) a single drop is placed in the eye or the nostril of each bird.

Oral instillation is a technique only used for avian encephalomyelitis vaccination. Here the contents of a full dropper are squeezed into the bird's mouth.

Direct injection

Certain diseases, Marek's being a prime example, require direct intra-muscular injection. Hand-gun vaccinators are obtainable, which deliver a measured dosage with each pull of the trigger. With the vaccine fed in from a container hundreds of injections can be made between refills. This is important in hatcheries, where thousands of chicks must be handled as quickly as possible. Recently, machine vaccination has been introduced, chicks being held against an opening through which a needle automatically administers their dosage. Trained staff are employed at hatcheries and while the skill is easily acquired, the poultry-keeper should attend a practical demonstration before attempting to inject birds.

Follicular and wing-web technique

Fowl pox/diphtheria is a disease which gains entrance to the body through tiny abrasions in the skin and the vaccination has to take a similar route, either via the feather follicles—for birds over ten weeks old—or wing-web stabbing.

In the follicular method about fifteen feathers are removed from the front of the shank, by upward plucking, care being taken to ensure that the follicles do not bleed. Vaccine dissolved in the solvent provided is then scrubbed into the follicles with a special brush. Reaction, after six to eight days, should show up as slight swelling of the follicles, occasion-ally accompanied by harmless encrustations.

The standard vaccination procedure at about thirteen weeks, where a unit has a history of fowl-pox, is the wing-web method. The wing-web of each bird is stabbed with a special needle—usually one with two prongs—previously dipped into a concentrated vaccine solution. The frontal web close to the shoulder is slightly stretched and the needle inserted through the skin, away from muscles, joints or blood-vessels. Localized reaction in a little over a week can be detected as small nodules on the skin. This system works well for chickens and turkeys.

Bacterial infections

Many bacterial problems in poultry are now well controlled by anti-biotics, improved hygiene and/or, in the case of diseases transmitted vertically from parents to progeny, by isolation and selective breeding, but the poultry-keeper cannot afford to drop his guard.

Salmonella

Most people associate the name salmonella with food poisoning and they are right, but that is a slight over-simplification. Bacteria often form very large 'families' of serotypes and the salmonella complex is no exception. So far about 1,800 varieties of salmonella have been identified. Many of these are quite harmless, but the 'family' also includes *S. typhi*, the cause of typhoid in humans, and serotypes like *S. typhimurium* which leads to food poisoning.

A number of salmonella serotypes are zoonotic, which means the same organisms can infect humans and birds or animals. This does not always apply; human typhoid is not apparently transmissible to other mammals for example, and certain salmonella diseases found in poultry such as pullorum and gallinarum, described below, are only clinically important in those species. Nevertheless, the existence of organisms which can cause disease in both livestock and man must be taken very seriously. The generic term for this type of infection, where it affects poultry, is avian salmonellosis.

Salmonellae can get into the 'food chain' at a number of points. They create a problem because birds and mammals (including humans) surviving an attack become long-term carriers of the infection. The symptoms may be no more than a minor stomach upset, sometimes not even that, but thereafter the carrier excretes salmonellae.

If breeding birds are infected, the organisms may reach the shells of their eggs before they are laid, or be picked up from contaminated nest materials. Occasionally, through ovarian infection, they may start inside the egg. This is a common route with ducks. Equally, however, contamination may result from handling of the eggs by an unwitting human carrier at the hatchery and this is one reason for the strong emphasis hatcheries place on hygiene, sometimes including medical checks on personnel. Salmonellae thrive in warm, humid incubator conditions. They penetrate shells, lowering hatchability, and spread among the chicks that do hatch.

Symptoms and deaths from salmonellosis in chickens, ducks and turkey poults tend to be confined to birds under three weeks of age. Often signs are vague and non-specific. Survivors then make a new reservoir of infection. Under the Zoonoses Order (1975) applied in Britain, if a veterinary inspection produces evidence of salmonella (usually this is verified by sending food or faeces samples to the nearest of twenty-four Veterinary Investigation Centres in England and Wales) it must be notified to the Ministry of Agriculture's Central Veterinary Laboratory at Weybridge, stating whether fowl, turkeys, geese, ducks, guinea-fowl, quail or other livestock were involved. In this way a record of salmonella in England and Wales is built up.

Assuming birds are not infected at the hatching stage they can come into contact with salmonella through their feed, if it includes contaminated ingredients. Fishmeal or other protein meals—especially those containing feather or meat-and-bone from poultry—are potential disease sources, but professional compounders now regularly monitor their feeds, and countries are tightening up their control of imported ingredients. At the beginning of 1978, for instance, a Protein Processing Order and an Importation of Processed Protein Order were both envisaged in British legislation and countries like Denmark already had strict control of imported animal foods.

Often, however, the damage is done when the birds are out of the stockman's hands. They can be infected as carcasses. It is easy to get the dangers of food poisoning out of perspective. A report by a Study Group of the British Association for the Advancement of Science issued in 1978 called 'Salmonella, the food poisoner' speculated, very cautiously, that the chance of being infected might arise from one meal in 20,000. It could be even smaller. Nevertheless poisoning does occasionally happen, nearly always through careless handling or cooking.

There are three main rules for safety.

a) Be hygienic. All food handlers and their working areas must be clean.

b) Cook properly. Salmonellae are killed at cooking temperatures, but frozen birds must be thawed to ensure that normal roasting time will bring the temperature inside to the same level as that outside the carcass.

c) Eat promptly. Meat left to cool slowly at room temperature is hazardous, especially if exposed to surfaces or utensils previously used for raw meat. Cooked meat can be safely frozen, but again, reheating must be to proper cooking temperatures throughout, and not just to a comfortable heat for bugs.

Salmonella pullorum

At one time a widespread threat to chickens, turkeys, ducks and guinea-fowl, pullorum is now far less of a menace thanks to energetic eradication under national poultry health schemes on both sides of the Atlantic, using blood testing techniques. It is often referred to in older textbooks as bacillary white diarrhoea (BWD), but this symptom is not always seen in infected birds.

Unhygienic surroundings, contaminated water and feed can all harbour the organism, but it is also egg transmissible. Symptoms in recently hatched chicks of huddling, cheeping, lack of appetite and occasionally pasty white diarrhoea around the vent, accompanied by varied mortality, are characteristic of pullorum. Early deaths denote trouble at the incubation stage but a mortality peak after the first fortnight suggests a brooder infection. Although symptoms are rarely seen after six weeks of age, it can also affect adult birds. Survivors frequently become carriers, the bacteria lodging in the ovary, from which they will infect future eggs laid by the bird. Various of the sulfa drugs, such as Sulfa Methazine and Sulfaquinoxaline are effective if used soon enough, but survivors will still be carriers.

Treatment, if necessary in a laying flock, is by inclusion of the drug furazolidone in the feed for ten days, but if breeding birds are infected they must be destroyed. This is where the agglutination blood test has been so effective in eliminating carriers. A drop of blood taken by pricking the wing vein is mixed with stained antigen on a testing plate. Positive reactors are quickly revealed by the blood granulating. Systematic testing in this way has reduced the incidence of pullorum so that now it is only applied to grandparent breeding flocks in the UK.

Salmonella gallinarum

Chickens, ducks, guinea-fowl and turkeys are all subject to gallinarum, which is otherwise known as fowl typhoid. It has particularly affected turkey flocks in America. The disease is serologically identical to *S. pullorum* and it reacts in the same way to blood testing, but *S. gallinarum* is more likely to be seen in growing pullets from about twelve weeks of age onwards.

An acute attack may be marked by sudden deaths, sometimes up to a third of a flock being lost. Other birds show increased thirst, depression and anaemia, with pale wattles and head parts. Body temperature and respiration increase and a watery yellow diarrhoea is evident.

As with *S. pullorum*, survivors carry the disease and pass it in their faeces as well as via their eggs. Where an outbreak is diagnosed, all reactors to the blood test must be destroyed. Furazolidone treatment in the feed can keep down the number of carriers. Birds should be moved to unpolluted surroundings. It is also possible to vaccinate against *S. gallinarum* where the risk is considered high.

Arizona disease

Serious losses have been recorded in American turkey flocks from Arizona disease, caused by another member of the salmonella group, but it has only once been detected in Britain, when a batch of quarantined birds in Wales were found to be extreting the organism.

This infection results in deaths among poults in their first weeks. General depression and nervous symptoms, with twisted necks, paralysis and diarrhoea, are common signs, and some birds may become blind, with pus formation in the eyes.

There is some danger for man, because this is a zoonosis which can lead to gastroenteritis, pneumonia and other ailments in humans.

Birds recovering from the disease are liable to be carriers and it can be transmitted to poults through the egg, or by external contamination. Furazolidone in the feed or furaltadone in the water is the recommended treatment (200 g/ton for five to seven days then 100 g/ton for one or two weeks) but control ultimately depends on elimination of carriers from breeding flocks, clean surroundings and high standards of hygiene in feed preparation, hatcheries and houses.

Avian Mycoplasmosis

When buying certain breeds of turkeys or chickens it is not uncommon to find them referred to as 'MG-free' or 'MS-free'. The latest addition to these cryptic notes is 'MM-free' which applies, at the time of writing, to just one strain of turkey, bred in England.

The initials stand for *Mycoplasma gallisepticum*, *Mycoplasma synoviae* and *Mycoplasma meleagridis*, three conditions produced by avian mycoplasma, the smallest free living organisms, also given the rather ponderous name pleuro-pneumonia-like organisms (PPLO).

By patient manipulation of breeding flocks it is possible to produce 'clean' birds, free of these infections, but it is an expensive and painstaking process and breeders make it a selling point when they achieve it.

Mycoplasma organisms, in themselves, do not necessarily cause losses, but they tend to be latent in stock which have not been cleared of the infection. Then their interaction with other agents, particularly ND, IB and *E. Coli* can flare up into serious problems. Their effect on performance, lowering egg output, impairing feed conversion, reducing growth rates and suppressing hatchability gives them an important economic influence.

Mycoplasma gallisepticum is primarily a respiratory disease of chickens and turkeys. It may be seen in the form of coryza, with runny eyes and nostrils, coughing and sneezing, or as infectious sinusitis in turkeys. The cheeks puff up grotesquely with mucus on one or both sides of the face. Internally, the lungs, respiratory tract and air sacs may be involved. Very occasionally it can lead to lameness and swollen hock-joints in chickens. A mild disease on its own, it frequently shows symptoms only when complicated by other infections.

Treatment is mainly by drugs and antibiotics administered in the food or water. This alleviates the symptoms. Total prevention is more complex. The disease can spread horizontally, from bird to bird, or vertically, to the embryo, via infection of the ovary and oviduct. Control therefore centres on identifying affected flocks, eliminating carriers among breeding birds and eradicating infection in eggs.

The egg treatment consists of either dipping pre-incubated eggs in a medicated solution—which must be cooler than the eggs to be drawn through the shell—or in heating them to 45°C (113° F) over eleven to fourteen hours, which runs the risk of lowered hatchability.

Mycoplasma synoviae is mainly associated with lameness and joint disorders in broilers and turkeys (infectious synovitis) but, in common with *M. gallisepticum*, it can cause respiratory symptoms, including air sacculitis and sinusitis. Swelling of the joints is common and the resultant difficulty in moving partly accounts for poor growth rates, although diarrhoea and anaemia are also symptomatic.

This organism spreads horizontally and vertically and is controlled by similar methods to *M. gallisepticum*.

Mycoplasma meleagridis is specific to turkeys. It can stunt their growth, cause perosis or slipping of the leg tendon in some birds and lead to abnormal wing feathering, crooked necks and air sacculitis, especially among poults.

Transmission of *M. meleagridis* from bird to bird can be through the respiratory tract, as with other mycoplasma infections, but it can also reach the oviduct venereally, from infected males, and in this way contaminate the eggs. Control by drugs, antibiotics and breeding techniques follows similar lines to the other mycoplasma treatments, but particular attention is paid to the hygiene status of the males before mating or artificial insemination.

Coliform infections

Many members of the *Escherichia coli* family of bacteria are to be found in the gut of normal animals and birds, living in complete harmony with their hosts, but the family does have its 'wild' serotypes which can cause trouble.

Coli septicaemia, a problem particularly noted with broilers, but also found in ducks and turkeys, often follows respiratory or other diseases. It can show up as a complication associated with mycoplasma infections for instance. Stress, as with over-stocking a broiler house, and debility trigger it off. Victims quickly lose interest in food and water and become lethargic. Occasionally respiratory symptoms are shown and air sacculitis occurs. Where commercial birds are involved, slow weight-gain and down-grading at slaughter cost the producer money.

Hygiene, stress-free surroundings and careful selection of stock—mycoplasma-free if possible—help to reduce the risk of this disease. Treatment is generally by furazolidone or furaltadone in feed or water respectively or by the use of tetracyclines.

Arthritis, with the development of pus in the hock joint, can be the result of *E. coli* infection.

Osteomyelitis can arise in developing birds when *E. coli* invades the bones. It sets up abscesses in the bone cavity and pus may form to swell the joints. Where the legs are affected, lameness is a feature.

Avian pasteurellosis

Within the pasteurella group of bacteria, several types may cause disease in poultry, of which one is particularly notable.

Fowl cholera arises from *Pasteurella multocida* and in its peracute form is a highly infectious and dangerous disease, particularly of waterfowl, though it also affects chickens and turkeys. Fortunately, in Great Britain at least, the peracute form is comparatively rare, although the disease has a worldwide distribution.

Peracute fowl cholera strikes fast and the first an owner may know about it is the discovery of dead birds in apparently good condition. Turkeys may show darkening of the head (cyanosis), other symptoms being refusal to eat, high temperature, a mucous discharge from the beak and green, evil-smelling diarrhoea. Mortality is high.

A chronic form of fowl cholera shows fairly typical bacterial effects, with respiratory distress, irritation of the eyes and lameness caused by joint lesions. Birds may also suffer infection of the middle ear.

Sometimes abscesses form on the wattles or other exposed skin areas, where a mild form of the disease has gained access through an open wound.

Little can usually be done against peracute cholera, because of its speed, but vaccines have been prepared for subcutaneous injection at eight to ten weeks, repeated four to five weeks later.

Ornithosis

Although not a common disease in British poultry, ornithosis is an infection found throughout the world which has special importance as a zoonosis. It is caused by members of a group of organisms known as Chlamydia, which can lead to serious illness or death in man. In the parrot family the disease is called psittacosis. Legislation now requires a licence for the importation to the UK of psittacine birds (parrots, parrakeets, budgerigars and so on) under the Importation of Captive Birds Order 1976. Where authorities identify the disease they have the power under the Psittacosis or Ornithosis Order 1953 to isolate and detain affected birds or their contacts.

In Britain the disease occurred in two flocks of ducks in 1954 and an outbreak, which may have been spread by pigeons, struck a flock of turkeys in the West Midlands during 1976, which indicates its comparative rarity among poultry, but it has an economic effect in some parts of the USA and Europe.

Symptoms vary according to the severity of the outbreak, but they include, as well as drowsiness and lack of appetite in sick birds, diarrhoea, respiratory distress, a rapid drop in egg production in layers and death. Birds can remain carriers for long periods.

The organism is voided in droppings and may be transmitted by inhalation. Should a veterinary diagnosis be positive, special care will be needed by the stockman to avoid unnecessary exposure to contamination. Workmen in processing plants have on several occasions become victims in the US. Luckily this disease is well controlled by broad spectrum antibiotics. For poultry, the inclusion of chlortetracycline in feed is effective.

Carcass disposal

A final word in this chapter must go to the sad business of destroying the 'failures'. It is essential to dispose of all diseased carcasses absolutely, to prevent infection from finding its way back to fit birds.

Incineration is a complete answer, but a high temperature is needed to consume everything without smell or inconvenience. A variety of suitable oil and LP gas-fired incinerators are on the market, equipped with after-burners. Typical examples will dispose of 50–100 kg (1–2 cwt) of waste in two hours.

Carcasses can be buried, but this is laborious, does not always guarantee that infection will not later be dug up and makes demands on space.

The alternative is a disposal chamber, best made of concrete, although stout timber will serve, which takes the form of a box, standing on loose earth. It can be sunk in the earth as a pit, but there must be no chance of the site chosen draining down into adjacent waterways to cause contamination. Dimensions will vary according to the number of birds being handled, of course, but around $2\,m^3$ ($2\,yd^3$) is adequate for a 1,000 bird flock. One essential is that the structure is well sealed. A small opening 305–380 mm (12–15 in) in diameter gives access at the top and must have a tight fitting lid to contain unpleasant smells as bacteria break down the organic material.

13 The egg business

The idea of making money from eggs has no special mystique about it. Plenty of youngsters, some not even in their teens, have abandoned paper-rounds when they have found they could get a better return from a few hens, and many older flock-owners extend the pleasure of their hobby by selling some of their output to local customers.

A world of difference may seem to separate this type of operation from the commercial producer depending on eggs for his livelihood, yet there are similarities. To make money in a small way calls for good stockmanship, minimum costs and customer service. Precisely the same principles apply to the full-scale commercial operation, only here the income must pay a living wage for the staff as well as the rental and other costs not normally incurred by the amateur. Anybody who wants to change their pleasurable hobby into a business must be prepared for a different attitude and a host of new problems.

If there is any secret to success in commerical egg production it can probably be summed up in one word—organization. Those who drift into the industry usually drift out again after a short time, complaining bitterly about their ill luck. The best producer can encounter setbacks, and if these occur early, before the business has found its feet, there may be no recovery. But the chances of survival are greatly enhanced by an organized approach. Aim to get quickly into profit and keep expansion within your available resources.

One of the first questions is naturally how big the business should be. It is very difficult to be categorical about this, since levels of efficiency vary and a flock total which supplies adequate profits in one case may barely scrape a living for somebody else. Many producers build up gradually from small beginnings, a hobby becoming a part-time occupation and progressing to a complete commercial enterprise. In this way experience and resources are acquired together. To begin with, houses and equipment are likely to be whatever is most easily obtained. The small producer may buy up secondhand cages at auctions and renovate them himself. Housing may be any convenient roofed area and ventilation 'natural'.

Needless to say, such a set-up makes big demands on the skill and ability of the poultry-keeper, who finds himself having to feed several thousand birds and collect the eggs by hand, as well as contending with temperature fluctuations and non-automatic manure disposal. But financially it sets the beginner off on the right lines, without the millstone of heavy repayments for new housing and equipment. It also emphasizes the important role of stockmanship. Good, purpose-built housing and up-to-date equipment certainly relieves the labour situation, but more birds per man does not necessarily mean more eggs per bird.

If the purpose of modernization is simply to improve efficiency rather than greatly extend the flock it may be found that the introduction of spray-on insulation to existing housing, improvements in light-proofing or the lining of leaky food troughs with a plastic liner will show more in the profit line than the purchase of the latest cage equipment.

When to start

A great many questions arise when considering egg selling, and time spent discussing requirements and prospects with experienced commercial producers and industry advisers before making a move can be invaluable in avoiding early difficulties.

All over the world, where any organized egg marketing exists, egg prices rise and fall on a cyclical pattern, in accordance with supply and demand. Strenuous efforts have been made by various marketing boards and agencies to iron out the price slumps, by imposing production quotas on egg producers for example, but despite these efforts, and the increasing 'outside' effects of imports and exports, the switchback is still there.

Although the complexities of today's market blur the picture more than in the past, it is still possible to predict likely over-supply periods by reference to the number of pullet chicks coming forward from hatcheries. These figures are regularly published in poultry journals and widely distributed by government sources, the International Egg Commission, Agriculture House, 25/31 Knightsbridge, London SW1X 7NJ, and, in the UK, by the Eggs Authority.

Since ultimate profitability depends on the margin of egg prices over the cost of the main inputs—feed, stock, capital and labour—clearly it is asking for trouble to embark on a business which is going to produce eggs just when there is a glut and prices are low. Established producers have no option but to ride out the bad times as best they can. The newcomer can wait if the omens are bad. Ideally feed prices should be stable or declining, while the chances of high egg demand when the flock are in full production should be good.

Market prospects

Obviously, without a market there is no business, so an important preliminary to setting up production is to investigate the selling opportunities. Consider the following questions:

What do people currently pay for eggs in the area?
Is their preference for white or brown?
What sizes do they mainly buy?
Do they pay a premium for brown eggs?
Would they pay more for personal service, freshness, stronger shells or some other plus factor?
How widely dispersed is the market?
How many customers can be conveniently reached?
Would there be a need for local advertising?
Could you sell to retailers?
Is there scope for 'farm-gate' sales—even a farm shop?
Are there any commercial packing stations in the locality willing to take all or part of the egg output regularly? On what terms?
What is the existing competition?

Answers to these questions do more than paint a picture of the market to be reached. They give a good deal of information about the business as a whole. They tell you, for example, the type of stock to be bought. The birds must be able to produce plentiful eggs in the optimum weight band, of the right colour and quality to suit customer demand. This narrows the field when checking through laying test results and breeder specifications such as those in Chapter 3.

Decisions about the market help to determine how big a flock may be and the prospects for future expansion. They also suggest the type of equipment needed. For example, distribution on a retail round or to local shops will call for packing and grading machinery, a stock of prepacks (the fibre or plastic boxes in which eggs are sold, usually in sixes and dozens) and suitable transport, even if it is only a bike and trailer for the really small-scale producer.

Supply of all eggs, or a proportion of them, to a local packing station does away with grading and packing on the farm, and the marketing responsibilities are removed, but naturally the potential rewards are less, allowing for the packing station's charges.

Finance and contract schemes

Every business is different in its cash requirements. The establishment of a new poultry production unit from the ground up, with full equipment, represents a major investment these days, out of the reach of most individuals. The birds and feed alone take a substantial sum, with layers' rations generally being very high in cost. It is possible to keep down the financial burden by gradual expansion and improvisation however, as has been seen, and fortunately there are various ways of obtaining the necessary backing. Banks, in particular, have become very conscious of agricultural needs in recent years. Many of them have set up special departments offering a considerable fund of advice on available grants and services. Some run schemes specifically for the poultry-keeper, providing loans for the purchase of buildings and fixed equipment and/or point of lay pullets.

Land purchase or large developments involving long-term borrowing are the province of finance companies, represented in the UK by bodies like the Agricultural Mortgage Corporation, building societies and merchant banks. Short- and medium-term loans of up to ten years are more the interest of the high-street banks.

Another potential source of finance for the poultryman is the commercial contract scheme. There are strings attached to contract schemes. They are run by large companies with something to sell—usually feed or chicks—but the producer derives benefits from the release of working capital which would otherwise be set aside for the purchase of food and birds.

In the British Isles the majority of schemes are run by national feed compounders such as BOCM-Silcock, Rank Hovis McDougall (RHM Poultry Bank), Spillers and Dalgety-Crosfields. The odd one out is the Sterling Contract Egg Production Scheme, now known as Ross Poultry Plans (RPP), run by Ross Poultry Great Britain Limited.

The general principle is that the producer signs a contract with a company, under which he agrees, usually for the period of one laying season initially, to run his unit according to the company's terms. Naturally, if it is a feed company, the terms will involve the use of their feeds and, if a chick company, the use of their stock, but since the products are normally of high commerical standard this need be no disadvantage.

Although details vary between contracts, they have a similar format. Started pullets, veterinary services and feed are supplied without charge at the outset, and a management fee to cover day-to-day running costs is paid to the producer weekly. A flock account is kept at head office and all the debit items, including a small figure for interest, are charged to it. One of the advantages pointed out by a contract producer recently, in fact, was that his paper-work was done for him. Credits, such as egg receipts and spent hen payments are balanced in the account and on depletion of the flock any credit balance is paid in full to the producer. If a debit balance is shown, a loss sharing arrangement may be implemented.

Among alternatives to the arrangement described is a rearing scheme in which chicks are supplied.

A criticism sometimes levelled at companies running contract schemes is that they encourage the inefficient to enter the industry, or help them to stay in it. This may, on occasion, have been true, but in effect the conditions laid down are quite stringent now. The poultry-keeper must be credit-worthy and be able to provide flock accommodation that meets the approval of the company. He will also have to conform to management standards such as the all-in/all-out system and be prepared to accept advice from the company's fieldsman. Performance levels yielding a hen-housed average of around 260-270 are looked for.

As a means of getting quickly into production and enjoying some of the benefits of the large organization there is something to be said for contract schemes. Those who are attracted by the idea should study the terms with real care before signing on, however. British producers can find help in this direction from the National Farmers Union, whose experts have issued certificates of approval to contracts whose terms 'strike a fair balance between the interests of the respective parties'. The NFU's sound advice is to ensure that you clearly understand how a scheme operates, the financial ties involved and for what you are responsible, both legally and financially.

Production and profit

It is often said that egg yield—the number of eggs per bird—is the leading factor determining margin of profit. That is not the complete answer, otherwise ducks would rule the roost, with their far more prolific output. Questions of public taste, egg size, feed consumption and even publicity also come into it.

Unfortunately for ducks, their value as egg layers is only truly appreciated in countries such as India. Nevertheless, returning to hen eggs, it is clear that the producer who obtains 280 eggs from each bird housed has an advantage over the one who only obtains 250, other things being equal. The 'inequalities' in such things as feed and pullet costs, labour and egg prices are the factors which can tip the balance between profit and loss and make the running of an egg unit a matter of fine tuning, where small adjustments can have a large effect on final results.

The successful producer is constantly concerned about keeping down mortality and feed wastage, limiting egg breakages, improving handling techniques and generally maintaining efficiency in a multitude of small ways. Some come with experience, others can be calculated systematically.

Food costs vary, but it can broadly be reckoned that they represent rather more than 70% of total egg production costs in the UK. Nobody, sensibly, pays more than he has to for feed therefore, but price-cutting must never be at the expense of performance.

Feed is cheaper if delivered in bulk rather than bags, and most commercial units in developed countries have bulk bins today. Quantity discounts are also obtainable and co-operative arrangements with other buyers are worth considering in this connection. Home milling and mixing is also a practical method of keeping feed costs down, if farm-grown cereals are available, although the initial cost of milling equipment and the fuel to run it must be taken into account. A further option in some areas is the mobile milling and mixing service provided by certain specialist companies who bring their truck-mounted plant to the farm and prepare fresh feeds on the spot from ingredients supplied.

Some producers in the UK successfully blend dried poultry manure—which becomes quite an acceptable, friable material—into their stock rations at up to 10% inclusion rates (see Chapter 9), but this involves drying costs and is not permitted in every area, as has been found in the United States.

Price-reducing methods which have an adverse effect on food quality or presentation at any stage of a bird's life are a false economy. Deficiencies in the nutrition of the parents can be carried over to the pullet chicks and affect their subsequent laying performance.

The chicks themselves should receive a high quality starter feed for their first eight weeks, followed by a pullet grower diet to point of lay. During this second period sample birds must be weighed, with a close eye on the body weights recommended for their particular breed at eighteen weeks.

Feed and breeding companies are generally agreed nowadays that energy intake must be controlled at this stage to bring the birds into lay at the peak of fitness, not fatness, though opinions differ over how this is achieved. Reliable suppliers are always ready with help on the correct allocation of their feed and it makes sense to follow their instructions. Poultry-keepers who purchase ready-to-lay stock should still check with the rearer on his choice of feed and programme to ensure a smooth transition to the layers' feeding pattern.

From a business point of view, a high, early peak in egg production is very important. This is not to advocate precocity. Birds encouraged into lay before they have reached a mature body weight will simply produce undersize eggs. The object is to have the flock producing at the optimum for the breed, because days lost by birds being late into lay will never be recovered. In a relatively small commercial flock of 5,000 layers, averaging 113 g (4 oz) of feed/bird/day, total daily consumption is 567 kg (1,250 lb), or more than a ton every two days. With feed costs being very high, a lot of money is being spent if nothing is being returned in egg sales.

Although nutrition is not the sole influence on laying performance, it does provide a useful throttle and brake business control for the poultry-keeper. Over and above the birds' essential dietary needs for growth and body maintenance, they must also have protein and the right balance of amino acids, energy, calcium and phosphorus to produce the required egg mass (egg numbers × egg weight). The producer can alter the average egg size up or down by changing the daily protein intake of his hens. The aim is to obtain the widest possible margin for value of output over the cost of input. In other words, when market prices for large eggs are good, you want the maximum number of eggs in that category, but only if the quantity and cost of the feed required leaves the best margin. Sometimes the cost of adjusting egg weight upwards will not be justified, and sometimes it can be more easily achieved by other means. House temperatures may need to be reduced, for instance. As temperature increases it cuts feed intake, so affecting egg size, at the rate of about 1 gm per 3°C in the upper registers. Water lines may need to be checked, since reduced water intake appears to be associated with decreased egg size. Birds will drink more if the water is cool, incidentally. Ahemeral light patterns—cycles of light and dark which add up to more than twenty-four hours, such as 14 light and 13 dim (2 lux)—regulate the time of ovulation and oviposition to produce larger, but fewer eggs. The ahemeral system is effective, but a little exotic and difficult to administer.

Livestock and other costings

An early decision must be taken over whether to buy pullets at point of lay or rear your own replacements, but often this is quickly resolved by the facilities and inclination of the stockman. In the long run, with established rearing premises and brooding equipment, it may be marginally cheaper to rear your own. This gives positive control over the dieting, lighting and general condition of the pullets and cuts down moving stresses. On the other hand, rearing is a business in its own right which many egg producers prefer to leave to the experts.

Whatever the decision, pullet quality is of paramount importance to the profitability of an egg business. Inferior birds occupy the same space and eat, if anything, slightly more feed than good stock. Breed is not always the deciding factor on performance either. In experiments at the Ministry of Agriculture's research farm at Gleadthorpe, in England, using birds of the same hatch reared on different farms, discrepancies ranging up to 75 p in the margin of egg sales over feed cost were found when they were housed together. The worst suffered from higher mortality rates, lower egg weights, delays in reaching maturity or higher food consumption than their sisters.

The moral seems to be to select one's rearer with care. At the end of lay, birds are normally sold off to old-hen processors for the best price prevailing at the time and bird depreciation is calculated by deducting the carcass value from the purchase price or the cost of rearing to point of lay. Timing of a cull has an important potential effect on profitability, since keeping the birds beyond their most prolific phase leads to larger, but fewer eggs, with weaker shells. Commercial producers normally keep a flock for a twelve-month laying period, which is about optimum. Occasionally, however, if there is a strong demand for large eggs they may be moulted and kept on for a second period.

Labour is generally costed at about 5–8% of total production costs but in one-man or family operations it may not come into the balance-sheet realistically at all. Hours are worked but never counted. This is an understandable piece of self-delusion, but if figures are to mean anything, labour should be calculated properly. It will have to be if the business expands sufficiently to take on paid help from outside.

Employment costs have soared in recent years, forcing egg businesses to protect themselves with increasing mechanization, so that an individual stockman aided by timed automatic feeding, lighting and egg collection, electronically controlled ventilation, powered manure removal and even failure alarms which phone him at his home or office, can look after tens of thousands of birds. Such elaboration is expensive to buy, run and maintain, but it is justified, indeed may be the only way to run large economic units with 20,000 or more birds in a house. Hand-labour, if not completely impractical, would be disproportionately costly.

In the small or medium-sized business a conscientious employee can cover his wages handsomely by enabling more birds to be kept, generally helping running efficiency with checks against feed wastage, ensuring temperature/ventilation requirements are met, inspecting the flock and keeping daily records. He or she may also relieve pressure on the employer, allowing more time for general observation and planning of improvements in production or marketing.

Well-trained staff are a considerable asset and in England the Agricultural Training Board (Bourne House, 32–34 Beckenham Road, Beckenham, Kent BR3 4PB) organize courses in poultry management for both school leavers and established workers.

The rest of the production costs—as distinct from marketing expenses—make up a relatively small segment of the whole, yet they have a real effect on the profit margin. Some are fixed. Others depend to some extent on the efficiency of the unit. The astute manager watches electricity bills, for instance, and may question the degree of saving fluorescent lighting would give over his existing tungsten bulbs, or whether the fans are being maintained and operated at peak efficiency.

The following list will not be comprehensive for every farm, but it is a fair illustration of the type of costs encountered with an egg unit.

    Electricity
    Water
    Farm transport
    Hygiene (including end-of-crop cleaning)
    Manure disposal (though if sold this could be an asset)
    Repairs
    Building and equipment depreciation
    Insurance
    Sundries (including VAT, telephone, postage and accounting charges)

Paperwork

Every poultry-keeper likes to know the progress of his flock but paperwork is regarded by many professionals as an irksome chore interfering with the 'real work' of egg production. That may be so, but often the success or failure of a commercial business depends on the accuracy with which it is done. If record taking is made habitual it is not difficult.

Large, well lit charts in the poultry house are a good deal more readable than the back of an old envelope, and less likely to get lost.

Egg production can be expressed on a 'hen-day' basis, which gives the current output of the flock, but a figure commonly used is the 'hen-housed' production. This is the total egg output divided by the number of birds originally placed in the house and it shows the combined effect

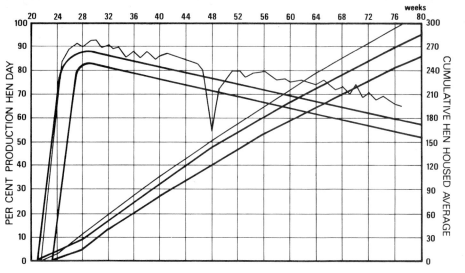

Performance record
This composite graph
shows the actual results of
a flock of nearly 4,000
layers which averaged 301
eggs hen housed to
seventy-seven weeks of
age. Bold lines show the
breeding company's target
performance and it can be
seen that the flock
achieved its good result
despite a fall between the
forty-seventh and fiftieth
weeks. Such charts
quickly show up problems
due to disease challenge
or management errors.
The 'knee-joint' line gives
hen-day production, the
diagram shows cumulative
eggs per hen housed.

of egg numbers and mortality. Examples of good in-house charts and
a typical 'knee-joint' graph with its ideal steep rise and gradual decline
in production, are shown here. Breeding companies produce these to
indicate the expected range of performance for their birds and filling
them in is an incentive to the producer to exceed their targets.

To calculate percentage hen-housed production the total eggs to date
are multiplied by 100, then divided by the number of pullets housed
times the number of days since housing. Percentage egg production per
week is arrived at by multiplying the total eggs by 100 and dividing
by the average number of pullets times seven.

For office use the producer must keep a note of the number and value
of the birds housed, a record of egg sales and sales of culls, details of
egg grading and seconds, the labour involved and the number and value
of the birds when sold at the end of lay. Daily recording in the house
of feed used and eggs collected can be usefully supplemented with data
on water intake and house temperature. These running figures can be
a useful early warning of potential trouble and a guide to any visiting
vet or adviser called in to check the flock. A simple inventory on feed,
light bulbs, egg trays and the like will also help to avoid those sudden
shortages which invariably happen when the supplier is out of reach.

14 The meat business

Is there any point in an individual poultry-keeper considering a start in meat production today? Surely the major producers have sewn up the market with the galaxy of poultry lines they now offer? For somebody new to the industry, that style of thinking is understandable, but not quite the whole story. True, if you check broiler prices in retail food cabinets and realistically account the cost of birds, feed, housing, heating, evisceration, packaging and transport—to name just a few items— you are unlikely to feel you can match the big suppliers, but do not be disheartened, other options do exist.

Broilers were the area of fast business growth in the 1950s and 1960s, but the pressures for high volume turnover meant that many companies flourished like hothouse flowers until they outgrew their economic strength and either collapsed or were integrated into yet larger food interests. As a result the modern broiler industry consists of relatively few, but massive combines which, with their economies of scale, present a formidable challenge to any newcomer. In other words, broiler production is not a very healthy business proposition, although a number of independent growers with their own housing and facilities supply the industry, under contract to one or other company.

On the other hand, turkeys, although also mass produced in frozen, oven-ready form, have retained their appeal as fresh, special occasion birds and are still sold as such by general farmers and others prepared to raise them in small quantities around Christmas. Oven-ready producers dislike this trade, which can disrupt an orderly market, and it is not always a money-spinner for the speculators, but it does point clearly to the fact that there are customers for lines which the big companies cannot, or will not, produce.

Enterprising small-scale poultry-keepers, therefore, look for something to distinguish their product, or give it a 'gourmet' tag. Quail and guinea-fowl come into this category, as do poussins, capons and geese. Some standard breeders have built a high reputation for traditional table fowl, reared to maturity. All these lines have specialized markets where they command good prices—essential when output volume is low.

Even finished layers or 'spent hens' are popular with some customers, who prefer them for flavour to young meat birds. In fact, until the 1930s spent hens made a contribution to table poultry supplies, alongside the spring chickens, capons, turkeys, geese and ducks grown for the purpose. Then came development of the broiler trade and with it the well-known trend towards cheap poultry meat for everyone. Finished hens were literally finished, except as ingredients for soups, pastes, certain prepared meals and pet-foods. Because of their high fat content they are a specialist job for processors, who pay by weight—and seldom enough in the view of commercial egg producers. Obviously, end of lay hens are no premium food item, although they make a very presentable coq-au-vin, and it would be impracticable to sell commercial flocks running into thousands at the farm gate, but certainly some producers have found a market for a proportion of their birds in this way. They are sold unplucked and unprocessed, direct to calling customers.

Those who buy special meat products are frequently quite enterprising themselves in pursuit of a bargain. Some like to see and select the birds alive, and are quite prepared to take them away for slaughter and evisceration, thus relieving the burden on the producer. At the other extreme, the producer may bring in customers from miles away by establishing a reputation for home-made pies or pâtés. If the taste and quality is right they will become staunch allies in recommending their discovery to their friends. This personal relationship is a strength the small producer should cultivate. Even pie-crusts monogrammed with the customer's own initials should not be beyond him if the price is right, where it would certainly be outside the scope of the mass production line.

Paradoxically, while the integrated company, by its nature, must make large, regular deliveries or 'drops' where they can be sold most quickly—usually a strictly defined range of stores or supermarkets—the small-scale producer has a wider choice of sales outlets.

Markets

Should you send your birds to market? At one time that would hardly have been questioned, but the pattern has changed to the extent that only big wholesale meat markets like London's Smithfield still exert any influence, and even their power has waned as the big poultry producers have taken over their own distribution and marketing.

The benefit to a producer of dealing through wholesalers is that it takes away his selling responsibility. Market dealers are professionals, in close touch with retailers, caterers and other buyers and will happily discuss their requirements with a supplier, but they drive a hard bargain and will expect the birds to be plucked, graded and packed. Some are specialists who have particular contacts with, say, the delicatessen trade and may find outlets for quail, guinea-fowl and poussins.

Caterers

It is worthwhile to spend time studying the local prospects and talking to caterers even before starting production. Some may be found willing to take oven-ready quail or guinea-fowl. Poussins, capons, ducks and geese may also interest them, but listen carefully to their terms of delivery, sizes, quantities and possible prices. If they have not thought of a line it may be possible to sow some ideas in their minds.

Portion control is important to caterers, and deboned chicken and turkey meat, rolled and packed, sausage-fashion, in fibrous casings, makes life simple for them. Such products are not beyond the scope of the small producer, but they are more likely to be supplied already by established producers, as are whole oven-ready chicken and turkey.

Remember that catering is not confined to hotels and restaurants. It takes in health farms, country clubs, company canteens, schools, hospitals, airlines, and while some may be out of reach, the horizon is broad.

Shops

Despite the arrival of hypermarkets, food shops still come in a wide variety of shapes and sizes, practically all with a chill counter and freezer cabinet. A selling point for the specialist poultry supplier is that the astute small shop-keeper is on the look-out for any line that gives him a plus over the supermarket up the road.

Delicatessen shops will sell quail and guinea-fowl, smoked meats, goose-liver pâtés, pies and similar products. Good butchers are likely to be interested in a full range of poultry, as are fishmongers and grocers.

If it is feasible, a 'farm shop' on the producer's premises offers several obvious advantages. It is immediately accessible, cuts out the middle man and gives the customer the sense of buying something absolutely farm fresh. Indeed, the keen poultry-keeper can use livestock as an added attraction, especially for youngsters. Some shops have been located so that birds can be seen moving in an adjacent house, through an observation window.

Private shops cost money to set up and run of course, and they have to be staffed regularly. Local advertising may also be needed initially.

Door-to-door deliveries

One way to become rapidly known in a district is to start a door-to-door delivery service. This is likely to be more expensive than other sales methods, taking into account the time and transport involved. It would be hard, if not impossible, to justify for any single meat line. On the other hand, if it can be combined with a regular egg round or something similar, it has prospects. Co-operation with others, such as dairy roundsmen, could be an answer, but the poultry-keeper must be confident of a fairly large and regular output.

Publicity, in the form of leaflets with suggested recipes can be easily circulated in advance using this system, with an invitation to customers to complete an order form and return it.

A further step in delivery services, successfully adopted by some, is the mobile farm shop consisting of a vehicle, preferably with the type of open side-hatch seen on ice-cream vans, which can be driven round the district.

Count-down to slaughter

Forward thinking is essential in poultry meat production if correct carcass weight and quality are to be achieved. Long before slaughter, therefore, the poultry-keeper should have a schedule which ensures the stock will be in peak condition at the time of marketing. Broiler companies have this worked out to a fine art, which is why they can be so categorical about killing ages.

In 1957 it took around seventy days for a meat strain chicken to reach 2 kg (4½ lb) liveweight. By 1977 it was taking fifty-two days or less, converting food to bodyweight at a ratio of 2:1. Through genetic and technical progress breeders can already predict that in another twenty years broilers could be reaching the same weight in thirty-six days with a food conversion ratio of 1.4:1.

There are, of course, individual variations, but broadly speaking, flock averages are predictable and weight can be adjusted down or up simply by altering the collection time. Having said that, however, there are differences in rate of growth and level of feed consumption between the sexes, and to achieve even greater flock uniformity and economy of production some growers separate them.

Thus, a flock of birds grown 'as hatched' may average 1.7 kg (3.8 lb) at seven weeks, FCR 1.92, but the males alone may average 1.9 kg (4.2 lb) at an FRC of 1.9, while the females weigh 1.5 kg (3.4 lb), FCR 1.94. It costs more to buy sexed chicks and house them separately, but those who can do it maintain that the extra efficiency more than compensates them. Either pullets or cockerels may be sold first, depending whether the processors want 'heavies' (cockerels grown on) and lighter weight pullets, or all birds at about the same weight, in which case the pullets will be kept a few days longer. Not only the feed but the floor space (0.5 sq ft for females, 0.65 sq ft for males) can be optimal under this system.

While the sex differences for broilers are relatively narrow, however, the contrast in turkeys is far greater. Market developments in England have led breeders to produce turkeys suited to three broad weight categories, each strain reaching its optimum at a different killing age. 'Mini' or lightweight birds, for whole bird purchase, kill out at about 4 kg (8.8 lb) at twelve weeks, when the males may weigh 4.5 kg (9½ lb) and the females 3.4 kg (7½ lb) liveweight. 'Midi' or middleweight turkeys, ranging up to some 7 kg (15¼ lb) at sixteen to seventeen weeks, will have stags (or toms) at that age weighing around 7.7 kg (17 lb) and hens about 5.7 kg (12¼ lb). The heavyweight strain for caterers and further processing, finds stags at 13.6 kg (30 lb) in twenty four weeks.

It is plain from these figures, that segregation of the sexes is desirable with turkeys. Because of their slower growth rate, hens may be fed lower protein, and therefore cheaper, feeds at certain stages of growth than the males. The birds also rear better in a less competitive atmosphere. Weights quoted here are not maxima, it should be noted. Breeding turkeys in England frequently attain more than 27 kg (60 lb) in a year's growth.

The perennial criticism of ducks by consumers is that they have too much fat and too thin a layer of meat spread over that broad keel bone, but breeders of modern commercial stock have worked hard to overcome this prejudice. In a six-year period Cherry Valley Farms, an important European duck company, claimed to have improved the amount of lean breast meat on their birds by 26%.

Growth rate can be phenomenal. A drake in a heavyweight breeding line from the same company, for example, has topped 5 kg (11 lb) in fifty-six days, though this is not a target for commercial producers of course.

Most fatteners produced for meat in Britain are now based on the Pekin breed, which has largely displaced the Aylesbury. It reaches a liveweight of some 3 kg (6½ lb) in seven weeks for a feed conversion of 3 : 1 or slightly better. Growers look for the stage known as 'first feather', when the flight feathers are around 80 mm (3 in) long and feathers have begun to fall from the neck, as an indication that the birds are prime for killing.

Geese have not had the degree of breeding refinement applied to the more popular lines of poultry, and since they are predominantly grazers, feed conversion is a matter for conjecture. The Embden × Toulouse, widely used for meat production in the UK, matures at twenty weeks of age, when the gander weighs approximately 9 kg (20 lb) and the goose 7 kg (15¼ lb).

Apart from their value as meat birds they are also kept to provide liver for pâté, and as much as 1 kg (2¼ lb) of prized feathers can be obtained from every five mature birds.

Geese need careful finishing and at one time were crammed, food being forced into them by hand or machine, twice daily in the ten days before slaughter. Such treatment would never be condoned by welfarists, although it is still practised in countries with a big export trade in goose livers, such as Israel.

Regardless of cramming, however, once the seasonal growth of grass has ended, birds retained can be more closely confined, preferably on straw litter in a convenient out-building with access to a pen for their last three or four weeks, and fed a supplementary diet of wheat, barley, oats and cracked beans. A mash including boiled potatoes, vegetable food scraps, animal protein, biscuit meal, flaked maize and skimmed milk, plus ample quantities of fresh greenstuffs will all help the finish.

Poussins and capons represent the two extremes of the chicken weight range. At the lower end, poussins are usually finished at around 1 kg ($2\frac{1}{4}$ lb) liveweight (650 gm or $1\frac{1}{4}$ lb eviscerated), although they can be lighter or heavier according to market demand. Broiler strains are sometimes used, but since the chick cost is spread over a very low bodyweight, surplus cockerels from laying strains are a frequent choice. These birds take about seven weeks to reach market weight, against five weeks for broiler strains and carcass quality may not be so good, but they are purchased cheaply from commercial hatcheries.

Feeding is normally by broiler starter crumbs, followed by broiler grower pellets and the birds may be housed either as broilers, on the floor or, more commonly, in cages.

Caponization is the feminizing of cockerels, or stag turkeys, to combine the advantages of the male's superior weight and the female's conformation and added fat. Originally it could only mean a surgical operation which involved an incision in the bird's side and the removal of the testicles using forceps. This is now forbidden in Great Britain and has been superseded by the much simpler chemical caponization, using synthetic oestrogens.

A syringe is used to implant hexoestrol or stilboestrol hormone (in the form of a 15 mg pellet or cream) just under the loose skin at the top of the neck. Manufacturers stipulate the minimum time for doing this before slaughter—usually around six weeks. It results in shrinkage of the testicles and reversal of the bird's sexual characteristics. The cock's comb does not materialize, it does not crow, becomes docile and a layer of subcutaneous fat is built up. After a period however, if the bird is not killed, the effects wear off.

There is no denying the succulence of capons, and repeat orders can be expected, but, particularly in America, reservations have been expressed about the possible consumption of medically harmful hormones. In effect the hormones are localized, and as additional precaution, the neck is cut off and thrown away.

Quail fatteners, kept in warm houses and fed on high protein turkey starter and grower diets, reach liveweights of 200–255 g (7–9 oz) in forty-two days. Feed conversion is about 3 : 1.

Guinea-fowl vary in their speed of growth according to the method of management. Kept extensively, in the open, the keets should reach market liveweight of just over 1 kg ($2\frac{1}{4}$ lb) in fourteen to sixteen weeks. Under broiler-style intensivism they achieve similar weights in ten weeks, at a feed conversion of 3 : 1.

A question of importance in the weeks before slaughter is whether stock have received any drugs or additives which need to be withdrawn. Caponizing compounds, as we have seen, come into this category. Cocci-diostats frequently have a withdrawal period of three to seven days, depending on type. Normally any requirements to cease a treatment before slaughter will be clearly indicated on the pack, but where there is the slightest doubt about additives or medicines consult your feed supplier or veterinary adviser and ensure that everybody concerned with the birds knows of the requirement.

After spending weeks bringing a flock to peak condition it is all too easy to throw the effort away by poor final handling.

Approximately twelve hours before slaughter—the precise time can be varied according to experience—birds should be deprived of food but not water. Starvation empties the crop and intestines and substantially reduces the mess and potential contamination during evisceration. It also has the inevitable effect of reducing the final weight by about 2% but this is more than offset by longer shelf-life and cleaner handling.

Weight loss from stressful heat, lack of water, long-distance haulage and general banging about, on the other hand, has no balancing advantages. Birds should be lifted carefully by the hocks, not the lower leg, placed, not thrown, into crates and protected where necessary from the elements. Unfortunately, if they are collected by processor's teams or private dealers, the poultry-keeper has no final control. The only thing to do then, if you are absolutely certain of your management and the condition of the birds when collected, is to watch the gradings and question any excessive losses through bruising, broken bones and deaths in transit.

Professional collection teams dealing with large flocks commonly work at night, when the stock are likely to be most docile, and the small producer can adopt the same procedure. Various loading and transporting methods are used, some employing loose plastic crates, others a system of fixed crates built up on the lorry. Sometimes conveyor belts are used on which the birds stand. The latest technique to be developed in England employs whole blocks of cages which are filled in the house, then carried by fork-lift truck to the lorry. Nearly 5,000 broilers an hour can be loaded by this method, which has also recently been adapted for turkeys.

Killing and preparation

While, at this stage, the broiler grower can turn to house cleaning and making ready for the next batch, the small-scale producer must, of course, continue with preparation for market. For this some basic equipment is needed.

Ideally a shed or room should be set aside for killing and plucking. The equipment in it will naturally vary according to the scale of operations, type of bird and form of marketing to be adopted, but basically, where preparation of oven-ready birds is envisaged on a regular basis, the inventory is likely to read as follows:

Electrical stunner
Bleeding cones and/or shackles
Dry plucker or scald tank and drum plucker
Sink with running water

Preparation table (washable surface)
Sterilizer
Knives, buckets/containers, cloths/paper towels
Packaging
Freezer

Later, if the size and scope of the business increases—say towards smoked products or sausages—the list will obviously extend.

Technique, rather than strength, is required to kill all but the biggest birds. Ducks, for example, because of their long necks, are sometimes said to be awkward, yet I have watched an expert kill them with no more than a twitch of the wrist. Grasping them two at a time by their necks as they put their heads out of a crate, and using their body weight, he had hung a dozen by their feet from a rack in little more time than it takes to read this sentence. That sort of unconventional expertise can only be learned by experience, and indeed, to avoid unnecessary suffering to the birds, first attempts should always be carried out in a calm, unemotional way, under the guidance of somebody with experience.

Chicken, guinea-fowl and turkeys up to medium size are normally not a problem, but quail must be handled carefully to avoid excessive damage to these small birds, while large geese and turkeys may need to be handled carefully to avoid excessive damage to the slaughterer.

The common method of killing small numbers of poultry is by dislocating the neck. Hold the legs of the bird firmly in one hand and the head between the first two fingers of the other, with the top of the head towards the palm of the hand. Pull the head downwards—without undue force, which can result in decapitation—turning the palm upwards simultaneously, and the neck is severed where it joins the head. Death is instantaneous, although nervous reaction will cause jerking and wing-flapping.

Large birds are more easily dealt with if hung in the shackles, while a technique sometimes recommended for the largest is to pinion the bird's wings, then place a broomstick across its neck, near the head. Hold the stick down with a foot at each end, and grasping the bird's feet, pull strongly and swiftly upwards.

Once killing has been accomplished the neck should be further stretched an inch or two. This is important because, with the bird suspended upside down, blood can drain into the cavity created, leaving the meat white. Failure to bleed birds leads to red, blotchy carcasses. In fact the large-scale commercial system of broiler and turkey slaughter consists of electrical stunning as the birds travel through on the moving shackles, followed by manual or mechanical cutting of the throat. The line then passes over a bath or trough to allow time for bleeding before the next stage of processing, and the blood is collected as a by-product.

In Britain, if bleeding is practised, stunning is required under the Slaughter of Poultry Act 1967, the only exception being for birds slaughtered under Jewish or Muslim ritual. Originally the Act applied only to chickens and turkeys, but it was extended in 1978 to cover guinea-fowl, ducks and geese.

Knife-type stunners can be obtained for small scale processing, which have an electrical blade operated by a switch on the knife handle. The current-carrying blade is placed against the bird's head. A properly stunned bird goes into a spasm in which the head is arched back and the tail drawn up. Rapid tremors pass through the wings. Contact should be maintained until these symptoms are seen, when the jugular vein can be opened with the knife, by cutting the side of the neck where it joins the head. There are other forms of hand-held stunner, such as electrified tongs, on the market, which some producers prefer.

After the vein has been opened the bird can remain suspended on shackles or placed head first in a bleeding cone. If the cut has been correctly made, blood will pump out and bleeding will have stopped in about one and a half minutes for a chicken, two minutes for a medium turkey.

Since feathers can be removed more easily while the carcass is still warm, plucking should not be delayed long.

Small batches of birds may not justify machine plucking, and plucking by hand, although laborious, does preserve the value of the feathers. Overalls, and advisably a face mask, should be worn for this dusty job. Feathers are pulled out in the opposite direction to which they grow, starting with the wing and tail feathers. Remaining pin feathers or stubs are best removed with the aid of a blunt knife.

Especially where ducks are concerned, small downy feathers should be kept in a sack or container separate from the larger feathers. Colours may also be segregated from white, depending on the potential market. Duck feathers currently fetch more, weight for weight, than duck meat.

Mechanized dry plucking cuts the effort and speeds the process considerably, but it still needs skill to achieve good results. A standard design developed in England in the 1930s and only slightly modified since, works on the principle of eight revolving steel plates driven by a $1\frac{1}{2}$-hp electric motor. Opening and closing at high speed, the plates remove the feathers and transfer them to a sack. A proficient operator will pluck sixty average chickens an hour with such a machine, more quail or guinea-fowl, and fewer turkeys or geese.

Wet plucking is the accepted method for all production line operations, scalding being necessary to loosen the feathers where upwards of 3,000 birds an hour are passing through the pluckers, but smaller commercial operators also use wet plucking extensively. If there is a difference it is in the temperature they use.

Broiler carcasses for frozen oven-ready production are 'hard scalded' at or above 56°C (138°F). For old hens, turkeys and ducks high temperatures are also used. Hard scalding reduces the bacteria count, acts faster and thereby reduces costs. Unfortunately it also tends to remove parts of the epidermis or outer skin, and birds allowed to dry out show this as brown patches. If immersed in cold water for a time, or frozen, the patches disappear and the method is satisfactory for birds sold frozen or going for further processing.

Fresh chilled birds, on the other hand, are given a 'soft scald' at 54° C (129° F) or less. Dip tanks with dial settings giving accuracies within a degree either way are made to the requirements of small processors, but any convenient tank and immersion heater arrangement is suitable providing it has reasonable thermostatic control.

Since dipping times and temperatures vary with the size and species being handled, trial and error must be applied with the first few birds. To ensure that water penetrates to the skin, carcasses are lowered head first into the tank and moved vigorously up and down. After, say, fifteen seconds, they can be tested by pulling at a few thigh feathers. If the feathers come out easily the scald time is sufficient. If not, further checks must be made every few seconds.

From the dip tank birds go to the plucker, usually of the 'drum' type, with a stainless steel bowl, into which they are placed. Rubber plucking fingers project from a revolving column in the centre, and this type of machine will pluck several carcasses at a time in a few seconds. Wet feathers are swilled away to the bottom for collection. A similar principle is applied in large-scale pluckers, but here the carcasses are carried by overhead conveyors between 'walls' of rotating fingers.

Carcasses removed from the plucker should be carefully scrutinized. A lot of skin blemishes will indicate too high a scald temperature; obstinate feathers suggest too short a dip time. Adjustments should be made accordingly.

Wax defeathering can be applied to a wide variety of poultry, including hens, but it is now usually confined to ducks and, sometimes, quail. After bleeding, the birds are immersed briefly in a bath of molten wax, then in cold water. The resultant coating is removed in one operation, leaving a cleanly stripped carcass. The wax is subsequently melted down for re-use and the feathers sieved out.

One advantage of wax plucking is the smooth finish it gives, but after plucking, by other methods, remove any remaining hairs or clinging down by singeing, either by gas jet, methylated spirits ignited in a dish or brown paper spills (reputed to cause less smoke than white).

Birds may be sold at this stage 'New York dressed', which means intact, with head and feet and with the guts still in. Sometimes a ruff of feathers is retained round the neck and hocks. In cool weather poultry will hang in this form for several days without deteriorating. Because of EEC disapproval it is a declining trade, however.

Once a bird is gutted, its shelf-life is shortened unless it is deep frozen. Fresh chilled birds must, therefore, be sold quickly. In large processing plants it is only a matter of minutes before the carcass has been eviscerated and brought down to a temperature of 4° C (39° F) either by air chilling or spin chilling in ice-cooled water, but the machinery involved is large, expensive and out of reach of the small-scale operator.

If oven-ready birds have to be kept for any period, therefore, a freezer cabinet or efficient chill room is essential. At least one producer successfully adapted a defunct freezer lorry for the purpose.

Preparing birds for table

To remove sinews

Make an incision at the back of the shank just above the foot and (1) prise up the sinews with a skewer.

Turn the skewer clockwise twice and bring the shank back to the thigh. Pull skewer away (2) and sinews will slip out. With turkeys use similar procedure but remove one sinew at a time.

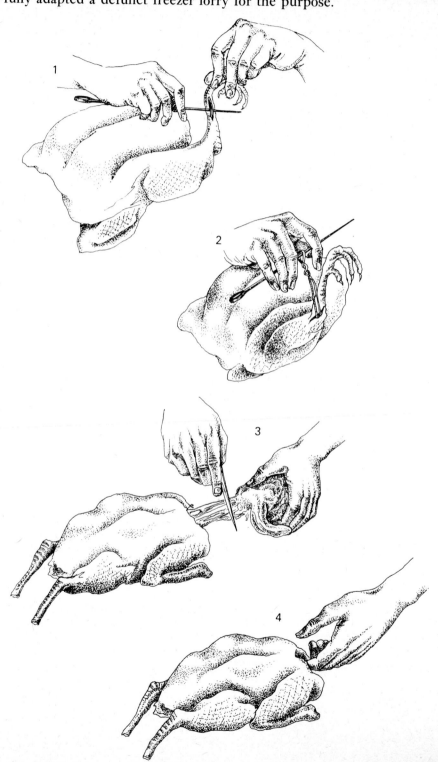

To remove neck

Lay carcass breast down, cut neck skin across with sharp knife about an inch above shoulders. Slit up neck, cut through muscle at base and break the neck away (3).

Peel away crop on inner side of neck skin, and cut out close to neck cavity.

Insert finger into neck cavity (4) and loosen heart by breaking away ligaments near breast bone. Then prise lungs away from their attachment.

Make a horizontal cut between the vent and the parson's nose so that the finger can be inserted to hook out the intestine. Cut round the vent taking care not to puncture the gut and remove with intestine attached (5).

Working through the vent cavity with fingers and thumb, free the abdominal fat and withdraw it with the gizzard (6). The viscera, heart, and lungs can then be removed, and finally the kidneys.

To truss

Fold neck skin over back and hold in place with wing clips. Place carcass on back and take string across thighs above hocks bringing it round and up alongside body (7) and lying across back (8). Finish by securing the wings.

Bibliography

ABC of Poultry Raising, J. H. Florea (Dover Publications, New York, 1975, republication of original by Greenberg Publisher, 1944)

American Poultry History 1823–1973, Editor J. L. Skinner (American Poultry Historical Society, Mount Morris, USA, 1974)

British Poultry Standards, Editor C. G. May (Butterworth, London, 1971)

Guinea Fowl, Van Hoesen (Stromberg Publishing, Fort Dodge, Iowa, USA, 1975)

Home Poultry Keeping, G. Eley (EP Publishing, Wakefield, England 1976)

Home Range Poultry Keeping, M. Gaisford (David & Charles Inc. (London) Ltd, 1978)

Keeping chickens, J. Walters and M. Parker (Pelham Books, London, 1976)

Keeping ducks, geese and turkeys, J. Walters and M. Parker (Pelham Books, London, 1976)

Management Guides 1–5, Poultry World/Poultry Industry (IPC Business Press, Sutton, England)

Modern Poultry Development, H. Easom Smith (Spur Publications, Liss, England, 1976)

Modern Poultry Keeping, J. Portsmouth (English Universities Press, London, 1965)

Poultry Production, L. E. Card and M. C. Nesheim (Lea & Febiger, Philadelphia, USA, 1972)

Practical Poultry Feeding, R. Feltwell and S. Fox (Faber & Faber, London, 1978)

The Histology of the Fowl, R. D. Hodges (Academic Press Inc. (London) Ltd, 1974)

Numerous bulletins of a very high standard are published by the Ministry of Agriculture, Fisheries and Food in England, written by ADAS specialists. All can be obtained through Her Majesty's Stationery Offices. Among those of special value to poultrymen are:

Climatic Environment of Poultry Houses Bulletin 212
Poultry Housing and Environment Bulletin 56
Poultry Nutrition Bulletin 174
Incubation & Hatchery Practice Bulletin 148
Intensive Poultry Management for Egg Production Bulletin 152
The Rearing of Pullets Bulletin 54
The Small Commercial Poultry Flock Bulletin 198

Glossary

ADAS Abbreviation for Agricultural Development and Advisory Service of the British Ministry of Agriculture, Fisheries and Food.

Aerobic Describes microbes requiring oxygen to develop and multiply.

AI Abbreviation for artificial insemination.

Air sacs Cavities containing air in birds to make them lighter.

Air space Also known as air cell. Between the shell membranes at the broad end of an egg, it provides the chick with air prior to hatching.

Albumen Main protein constituent of plant and animal tissues. The white of an egg is almost pure albumen.

Amino-acids The form in which proteins are absorbed into the blood-stream.

Anaerobic Microbes not requiring oxygen to develop and multiply.

Analysis Estimate of the composition of individual feeds, particularly in terms of minerals.

Antibody Agent formed in the liver, spleen and bone marrow and circulated in the blood as a defensive response to disease challenge.

Ark Portable range unit with slatted floor, for growers.

Autosexing Breeds in which sex can be determined at day-old by differing down colour of pullets and cockerels.

Axial feather Small feather separating secondary from primary feathers.

Balanced ration A diet containing all the necessary ingredients, in correct proportions, for body maintenance and healthy performance of stock.

Bantam Broadly refers to diminutive chickens, including small versions of standard size birds bred down, or created by judicious crossings between birds so produced. True bantams have no large counterparts.

Barring Alternative strips of light and dark across a feather.

Beard Feathers under a bird's throat, characteristic in Houdan, Faverolle etc.

Bib Another name for beard, or may refer to white breast marking in certain ducks.

Blast freezing Commercial freezing method employing refrigerated air blown over product in a tunnel or chamber.

Blood ring Denotes early embryonic death.

Blood spots Internal fault of eggs, caused by minor haemorrhages during their formation. A common cause of down-grading, detected by candling.

Boiler Boiling hen—an old bird requiring thorough cooking. Not to be confused with broiler.

Broiler Name, originally applied in America, for young chicken killed for meat. Eight weeks old or less.

Broody Bird with desire to sit on eggs.

Btu British thermal unit. It is the quantity of heat required to raise the temperature of 1 lb of water by 1°F.

Californian cage Layout in which tiers are staggered to allow droppings to fall clear to pit below.

Calorie Amount of heat needed to raise 1 g of

water by 1°C. The kilocalorie (Kcal) used for measurement of energy levels in nutrition is equal to 1,000 'small' calories (but see also Joules).

Candling Examination of eggs against a bright light source—originally a candle.

Cannibalism Killing and eating of birds by their fellows. Often starts by vent pecking and may result from too large a colony or other stress factors.

Capon Cockerel, feminized by surgery or chemicals to improve eating quality.

Caseous Cheese-like. Frequently used to describe pus conditions.

Cfm Cubic feet per minute. Measurement of air movement.

Checks American expression for cracked eggs.

Cholesterol Present in eggs and frowned on by some dieticians, but in fact a natural component of cell tissue—including the human brain.

Clears Incubated eggs showing no signs of embryonic development when candled.

Clubbed down Embryonic abnormality in which down appears beaded into small nodules.

Colony Collection of birds interacting with one another, whether in cages or larger flock conditions.

Compounder A manufacturer of prepared feeds.

Concentrates Additives to provide a ready-made protein/mineral/vitamin balance when mixed with cereals. Designed mainly to simplify on-farm milling and mixing, they have a protein content of 40–45% and are normally included at about 15%.

Contra-flow chiller Machine for cooling meat-birds, in which water travels in opposite direction to carcasses. Now specified by EEC regulations.

Cramming Old-fashioned method of fattening birds by force-feeding.

Cull To remove and kill ailing or surplus livestock.

Custom hatching Hatching for customers at an agreed price.

Cuticle The 'bloom' or residue formed on a new-laid egg as it dries.

Day-old Term to describe chicks and poults, normally packed and despatched within twenty-four hours of hatching. Allowing for spread of hatch and time of transport, however, it is a loose term.

Debeak Removal of part of the mandible to prevent feather-picking and cannibalism.

Deep litter Floor covering, so-called because it is allowed to build up depth with a succession of fresh layers. Nowadays litter is normally completely replaced between flocks.

DPK Domestic poultry-keeper.

DPL Dried poultry litter. A mixture of manure and floor material from floor-housed birds.

DPM Dried poultry manure. As obtained from caged birds.

Dubbing Removal of comb and wattles at 1–7 days, with curved scissors.

Ectoparasites External parasites such as fleas and mites.

Egg binding Blockage of oviduct by extra large egg, an egg broken internally, twisted or inflamed oviduct or other obstruction. May be helped by smearing with petroleum jelly and manipulation.

Egg tooth Horny protrusion on hatching bird's beak to help in breaking shell. It falls off later.

Elasticity of demand The amount of change in consumer income or product price needed to achieve a change in demand.

Endoparasites Internal parasites such as worms and coccidia.

Enzyme Agent which breaks down food in the gut.

EPEF European Production Efficiency Factor. A measure of broiler crop production efficiency.

Evisceration Removal of the internal organs in preparing poultry for eating.

FCR Food conversion ratio. Usually expressed as the ratio of food consumed to liveweight or weight of eggs produced. Thus two pounds of feed for one pound of liveweight is FCR 2:1, or just FCR 2.

Flats Egg trays.

Flights Main flight feathers, also called primaries.

Fowl In this book means chickens, although American usage embraces all poultry.

Free range Old-time system of allowing birds to range over wide area of land.

Gander Male goose.

Getaway cages Cages developed with nest boxes and perches, to allow layers more independence of movement.

Grain balancer Ration specially prepared for use with grain.

Green gosling Young goose fattened for Michaelmas, September 29.

Grower Pullet between six weeks and point of lay. Can also refer to poultry-keeper specializing in rearing birds.

Growers' mash Ration designed to bring growers to optimum condition at maturity.

Growth promoter Feed additive to enhance weight gain and optimize feed conversion.

Hackles Long, narrow feathers on neck and, in males, the saddle area of show birds.

Hair cracks Fine cracks on egg shells.

Hammer mill Feed mill employing 'hammers' to crush grain.

Hard feather Classification of show birds.

Haugh unit Measurement of height of egg white when broken out. Height denotes freshness and quality.

Hen housed average Total number of eggs laid by a flock, divided by the number of birds first housed. This takes into account both rate of lay and mortality and is a good indication of relative profitability.

Hock Joint between lower leg and thigh.

Hover Brooder with canopy placed low over chicks. Largely replaced now by suspended radiant brooders.

Hybrid Product of two or more pure lines.

Hybrid vigour Hardiness and superiority in certain characteristics of a hybrid to either of its parents.

Intensivism Generally refers to keeping birds in 'intensive', i.e. closed house, conditions.

Intermittent lighting Alternate light and dark periods used for intensively housed table poultry.

Joule Measurement of energy. In feeds, energy levels are frequently measured in megajoules/kilogram, a successor to Kcals/kg. Divide Kcals by 239 to find M joules.

Keel Breastbone or sternum of a bird.

Keet Young guinea-fowl.

Keyes tray Egg tray in fibre or plastic usually designed to hold thirty eggs. A case of eggs holds twelve trays totalling 360 eggs.

K-factor Amount of heat, in Btu's, which will pass through one inch of material in one hour, per square foot, for each degree F difference between the surfaces. That is Btu/in/ft²/hr/deg F.

Kosher trade Special trade according to Jewish ritual.

Lacing Feather marking in which edging is of a different colour to ground colour of feather. Where there is an additional inner band it is known as double lacing.

Lagoon Pit containing water into which manure is put to break down bacteriologically.

Least significant difference Smallest meaningful difference between two results.

Livability Ability of birds to survive to a given age.

Livestock depreciation Margin between return on finished hens and cost of replacements.

Lumen Measurement of light intensity. One lumen/ft² is the same as one foot candle.

Lux Lumen/m². One foot candle is 10.8 lux. This is roughly equivalent to 15 W tungsten filament bulbs at 3 m (10 ft) centres in a highly reflective (e.g. aluminium lined) house.

Mandible Hard parts of the beak or bill.

Mash Dry mash—milled and mixed feed ingredients—most widely used for poultry rations. Crumbs and pellets are commonly fed to meat birds.

ME Metabolizable energy. The level of available energy supplied by complete rations and ingredients. Once measured in Cals/lb, now quoted in M joules/kg.

Meat and bone meal Feed ingredient with 50% protein.

Meat extender Vegetable protein added to meat products.

Meat to bone ratio Balance between meat and bone on a carcass.

Micro-ingredient Additives included in rations in tiny amounts, e.g., growth promotants.

Morbidity Depressed condition through disease, not necessarily resulting in death.

Mortality Assessment of death rate.

Moulting Seasonal shedding of feathers by birds.

Mushy chick Colloquial description of chick with inflamed navel or omphalitis.

NYD New York Dressed. Form of carcass presentation in which bird is uneviscerated, with head and feet on.

Notifiable disease Disease which must be reported to the police or authorities.

Oöcyst The 'egg' stage of coccidiosis, passed out in the droppings to be picked up by other birds from the ground.

Oven-ready Plucked, eviscerated and prepared for the oven.

Packing station Usually a centre for egg packing and despatch, but can also be applied to poultry processing plant.

Paraformaldehyde prills Convenient form of formaldehyde concentrate for fumigation.

Pâte de foie gras Food delicacy made from goose-liver.

Pathogen A disease-forming organism.

Peck order Order of 'seniority' developed in a flock.

Pendulous crop Seen as sagging area at base of neck particularly in aging layers. Not serious.

Peristalsis Muscular wave action of gut.

Persistency Layer's ability to continue production over a prolonged period. An important economic factor.

Pfu Abbreviation for plaque forming unit, a measure of the potency of a vaccine.

Phenotype Outward appearance of a bird, rather than its genetic make-up.

Photoperiod Period of lighting.

Pin feathers Undeveloped feathers, forming short stubs.

Pinholes Holes in egg shells caused by bird's claws or beak.

Pinion Tip of bird's wing, from the last joint.

Pipping Act of breaking out of the shell.

Point of lay Age at which layers start to produce eggs. In chickens generally at about twenty to twenty-two weeks.

Pop hole Doorway through which poultry enter and leave a house.

Poult Young turkey.

Poussin Small chickens in liveweight range around 1 kg ($1\frac{1}{4}$–$2\frac{1}{2}$ lb) sold for specialist gourmet trade.

Precocity Start of lay before a pullet is physically prepared. Small and soft-shelled eggs usually result.

Premix Mixture of cereal base with small amounts of concentrated ingredients to ensure adequate blending with a manufactured ration.

Prepack Egg pack for retail sale and display.

Primaries Main flight feathers.

Progeny test Means of assessing parents by testing performance of their offspring.

Pterylae Feather tract.

Qualitative restriction Control of bird weight by regulating nutrient content.

Quantitative restriction Direct feed limitation to control bird weight.

Quartile ranking Ranking of bird performance by statistical analysis of results divided by averages.

Random sample test Comparison of strain performance by independent authorities selecting at random from stocks available.

Rachis Main shaft of a feather.

Ranikhet Indian name for fowl pest (Newcastle disease).

Replacements New stock, bought in to replace birds at the end of their productive life.

Roche scale Scale for measuring depth of yolk colour.

Roping Removal of lower part of intestines of meat birds to help their lasting qualities. Now outmoded by oven-ready trade.

Rose comb Style of comb flattened to head, covered with small nodules and finished with a spike or leader.

Saddle Part of back, from centre to tail, on cockerels. In females the corresponding area is called the cushion. Also applies to canvas cover looped over wings and across back of hen turkeys to protect them from spurs and claws of mating stags.

Schizont Part of life cycle of coccidia in the intestine.

Secondaries Inner quills on the wings, separated from the outer primaries by axial feather.

Self-colour Usually refers to a breed of one colour throughout, such as white or buff.

Semi-stepped cages Tiers of cages half overlapping to save house space.

Setter Initial stage incubator from which eggs are transferred to hatcher at time of pipping.

Sexer Person employed to separate male from female day-olds.

Sex-linkage Where progeny inherit characteristics of opposite parents—pullets from sires and cockerels from dams—making it possible in some instances to differentiate male from female chicks by down colour or feather development.

Sexual dimorphism Difference between average male and female performance within the same flock.

Shooting the red Age at which stag turkeys develop red caruncles around head and neck, about eight to ten weeks.

Sib Progeny of brother-sister mating.

Sickles Top pair of curved feathers in cockerel's tail.

Sided yolk Yolk offset from centre of egg.

Single comb 'Conventional' comb with single upright blade.

Snood Fleshy protruberance over beak of turkey.

Snow chilling Chilling carcasses by carbon dioxide 'snow'.

Spangling Effect produced when feathers are tipped in a different shade to the bird's ground colour.

Spraddle legs Spreading of legs and difficulty in standing of newly-hatched chicks. May be helped by paper or coarse-woven cloth in hatch trays.

Stag Common British term for male turkey.

Starve-outs Young birds, particularly poults, dying through failure to find food or water.

Strains Varieties of stock which will produce the same traits from generation to generation.

Straw-yard Method of keeping poultry in yard or enclosure on deep straw.

Stubs Partly grown feathers or remnants of plucked feathers. Rich in pigment, they can be a nuisance by staining the skin of dark-feathered meat birds.

Sulpha drugs Sulphonamides. Drugs widely used in control of coccidiosis and other ailments.

Supplement Vitamin and mineral reinforcements added to poultry rations.

Surfactant Surface active agent—description of certain hygiene chemicals.

Tail coverts Lesser tail feathers immediately below curved sickles on cockerel.

Toe punching Method of identifying breeding birds by punching holes in foot-webs.

Tom Common American term for male turkey.

Trap nest Nest which traps breeding females so that records of their eggs can be taken before they are released.

UGF Unidentified growth factor. Nutrients in various feed ingredients such as fishmeal, not positively identified but having a clear effect on performance.

USDA United States Department of Agriculture.

U-value Quantity of heat (in Btu's) which will pass through one square foot of a barrier in one hour to alter the temperature by one degree F. I.e. Btu/ft^2/hr/deg F. Metric U is Imperial U $\times$ 5.678.

Veranda Outdoor pen or coop having a slatted run area above ground level.

Vertical integration Integration of various stages of a business, from production to marketing.

Wing band Identification tab fitted to wings of breeding birds.

Xanthophyll Found in grass meal, yellow maize and fresh greenstuffs, this carotenoid pigment has a bearing on meat and egg-yolk colouring.

Zoonoses Diseases transmissible between man and animals.

Index

Page numbers in italic refer to illustrations